D0965829

THE SOLID STATE: X-RAY SPECTROSCOPY

THE SOLID STATE: X-RAY SPECTROSCOPY

L. JACOB D.Sc., Ph.D., F.Inst.P.
Research Associate in Physics,
University College, Dublin

A HALSTED PRESS BOOK

JOHN WILEY & SONS
New York – – Toronto

English edition first published in 1974 by Butterworth & Co
(Publishers) Ltd
88 Kingsway, London WC2B 6AB

Published in the U.S.A. and Canada by Halsted Press, a Division of John
Wiley & Sons, Inc.,
New York

Library of Congress Cataloging in Publication Data

Jacob, L.
The solid state: x-ray spectroscopy.

"A Halsted Press book."
Bibliography: p.
1. Energy-band theory of solids. 2. X-ray spectroscopy. I. Title. [DNLM: 1. Spectrum analysis.
QC451 J15a 1974]
QC176.8.E4J32 530.4'1 74–5391
ISBN 0–470–43440–6

Text set in 10/12 pt IBM Press Roman, printed by photolithography,
and bound in Great Britain at The Pitman Press, Bath

PREFACE

The subject of the X-ray spectroscopy of the solid state has grown steadily over the past 30 years, and has provided the first direct experimental evidence for the existence of the Brillouin zones in a solid. The time therefore seems appropriate for setting down in book form an outline of one of the main experimental approaches to the problem of band structure in the solid state.

The purpose of the book is to present the evidence obtained by the soft X-ray method about the band forms and structure of some simple metals, compounds and alloys. It is addressed to students of Physics, Chemistry, Engineering, Metallurgy, Materials Science and Biology, and should serve as an introduction which is central and complementary to the band theory of the solid state.

In order to have the necessary background for understanding the basis of interpretation of soft X-ray spectra, the student is given a résumé of atomic structure and spectra, by way of introducing the concepts involved in appreciating the binding forces in solids. This is followed by a general account of the quantum theory of electron energy states and the implication it has in connection with the form and character of the bands. The soft X-ray technique provides an experimental method for checking theoretical predictions and is described in some detail. The selected examples given, of the resulting spectra for simple monovalent, divalent, trivalent metals, and some compounds, illustrate the power of the method; the interpretation of the band forms giving the density of occupied states is based on concepts brought out in the early chapters, and throws light on the direction and advances of band theory. Some of the features of absorption spectra which give information about the density of unoccupied states are also noted.

The central problems in investigations of this kind are concerned with band structural changes based on the simple models presented; these must necessarily be modified to take account of changes in environment of the valence electrons, for example, the formation of alloys. The final chapter is devoted to a review of both experiment and theory, and some consideration is given to the factors necessary to bring them more closely and fruitfully together. A brief account is also given of advances in the past ten years using the different approaches of 'electron spectroscopy' for unfolding valence band states and structure. The information these yield supplements and may extend our knowledge gained by the X-ray method; the next decade should provide interesting material in this field.

The author wishes to thank Messrs. Fabian, Sherwood and Little for their comments on the major part of the manuscript. He would like to regard this small effort as a tribute to the memory of the late H. W. B. Skinner who so brilliantly opened up the soft X-ray field of research.

L. Jacob.

CONTENTS

Publishers note: To avoid possible confusion between nu (ν) and an italic 'vee' (v) in the equations, velocity has been represented throughout this book by a lower case roman 'vee' (v).

Chapter One

RÉSUMÉ OF ATOMIC SPECTRA AND STRUCTURE

1.1 Introduction

Solids are formed by conglomerates of atoms held together by various types of inter-atomic forces which result from interactions between the outer or valence electrons of the atoms. We shall, therefore, expect some modification of the structure of the atomic energy levels of the individual atoms when these forces act to produce the bonding between atoms. In order to develop a spectroscopy of the solid state, it is hence necessary to look at the bases on which the energy level concept has been founded. This goes back to before the First World War, when Bohr advanced the idea of electrons in stationary states of the atom to explain the origin of the lines in the spectrum of atomic hydrogen.

1.2 Spectral Series

The lines in hydrogen spectra had been classified empirically by Balmer as far back as 1885 in a formula for the wave number $\bar{\nu}$:

$$\bar{\nu} = \frac{1}{\lambda} = R\left(\frac{1}{2^2} - \frac{1}{n^2}\right) \tag{1.1}$$

where R was a constant and n took integral values 3, 4, 5 . . .

The fact that the wave number of a spectral line could be expressed as the difference of two terms was developed by Ritz in 1908 into his combination principle. Not all differences obtained in this way yielded lines. No satisfactory physical interpretation of this existed until 1913, when Bohr, using the Rutherford planetary atom model, introduced quantum ideas to develop his theory of stationary energy states. In this, the electron rotates in a number of fixed orbits round the central proton without radiating energy; this hypothesis contravened classical electromagnetic theory, which placed no restriction on the number of orbits and predicted that the loss of energy of the rotating electron would cause it to spiral into the central nucleus. According to Bohr, radiation is emitted or absorbed in discrete amounts as quanta of energy, $h\nu$, *only* when the electron moves from one stationary energy state E_i to another E_f, that is, the energy change

$$E_i - E_f = h\nu \tag{1.2}$$

where h is Planck's constant and ν the frequency of the emitted radiation. When generalised, this means that all atoms have electrons in a series of stationary energy states and that quanta of energy are emitted or absorbed by transitions between such states. The term values thus fall into place, since:

$$\frac{E_i}{hc} - \frac{E_f}{hc} = \frac{\nu}{c} = \frac{1}{\lambda} = \bar{\nu} \tag{1.3}$$

and each term value represents an energy state. The question of allowable and forbidden transitions was settled by the later development of the vector model of the atom. The proposal of Uhlenbeck and Goudsmid in 1925, to allow for spin states, meant that spin–orbit coupling became a factor in determining the energy states; also it provided an explanation for the 'doublets' observed in the atomic spectra of the alkalis.

As all later work, particularly the interpretation of X-ray spectra leans heavily on the idea of energy levels put forward by Bohr, it will perhaps be in place here to derive these briefly according to his scheme for the hydrogen atom. It should be mentioned that the approach from the wave–mechanical side in which each electron state is represented by a wave function ψ satisfying the Schroedinger wave equation which, in one dimension, has the form:

$$\frac{d^2\psi}{dx^2} + \frac{8\pi^2 m}{h^2}(E_{total} - E_{pot})\,\psi = 0 \tag{1.4}$$

leads to the existence in the atom of discrete energy levels confirming the intuition of Bohr. In his application of classical mechanics, he equated the centrifugal force mv^2/r to the Coulomb attraction of a nucleus of charge Ze, i.e. $Ze^2/4\pi\epsilon_0 r$, and introduced the quantisation of angular momentum:

$$mvr = n\frac{h}{2\pi} \tag{1.5}$$

where n is a quantum number 1, 2, 3 . . .
The total energy, kinetic and potential is $\frac{1}{2}mv^2 - Ze^2/4\pi\epsilon_0 r$ and this yields on substitution:

$$r = \frac{n^2\epsilon_0 h^2}{\pi m e^2 Z} \tag{1.6}$$

$$v = \frac{e^2 Z}{2n\epsilon_0 h} \tag{1.7}$$

$$E = \frac{-me^4Z^2}{8\epsilon_0{}^2h^2n^2} = -\frac{RhcZ^2}{n^2} \tag{1.8}$$

where R, the Rydberg constant, $me^4/8\epsilon_0{}^2h^3c$, has the value $1.0974 \times 10^7 \mathrm{m}^{-1}$.
As

$$\nu = \frac{E_\mathrm{i}}{h} - \frac{E_\mathrm{f}}{h} = RcZ^2\left(\frac{1}{n_\mathrm{i}^2} - \frac{1}{n_\mathrm{f}^2}\right),$$

then

$$\bar{\nu} = RZ^2\left(\frac{1}{n_\mathrm{i}^2} - \frac{1}{n_\mathrm{f}^2}\right) \tag{1.9}$$

In this way he expressed $\bar{\nu}$, the wave number, as the difference of two terms, where for the Balmer series $n_\mathrm{i} = 2$, $n_\mathrm{f} = 3, 4$ . . . The value of R obtained from spectra agreed closely with that derived from the values of the constants after a slight correction to allow for the 'reduced' mass, since the electron and proton revolve round the common centre of mass.

It is clear that where we have a single electron attracted to a nucleus of unit positive charge as in hydrogen-like atoms, for example in singly ionised helium, He^+, or doubly ionised lithium, Li^{2+}, the energies involved vary as Z^2, i.e. 4, 9 respectively, the radii of orbits vary as n^2/Z and the velocity in the orbit varies as Z/n. An extension of Bohr energy concepts can be made for the heavier atoms; the valency or outer electrons are mostly screened from the nuclear charge by the core electrons, and in a sense are in a hydrogen-like field, so that a modification of Z by the introduction of a screening factor s gives a good representation for the energy levels.

A general scheme for a system of energy levels, $C\,Z^2/n^2$, based on the ideas of Bohr is shown in *Figure 1.1.* Here, the normal state $n = 1$ is taken as the zero or ground level and the higher energy states are shown by horizontal lines for the various principal quantum numbers n; the energy imparted to an electron in level $n = 1$ by electron or ion impact, or by absorption of radiation, may send it out to one of a series of higher energy states, $n = 2, 3 \ldots$ The final result is a population containing varying densities of excited electron states. In falling back to the ground state, these electrons may release the energy of excitation as discrete quanta of radiation of energy $h\nu$, either in a one-step process,

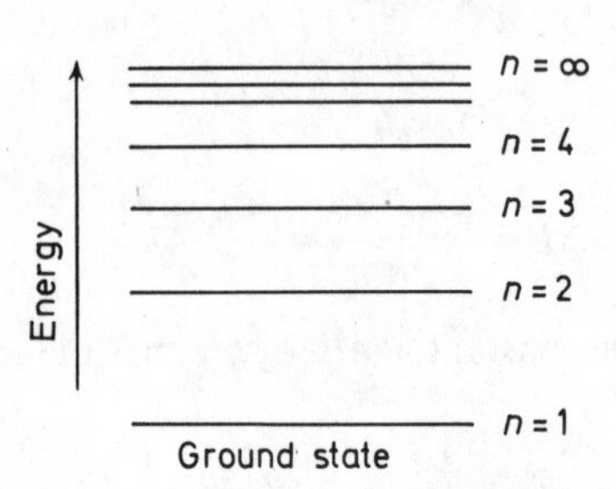

Figure 1.1 General scheme of atomic levels

or, in a series of step-like transitions between excited states, for each of which radiation quanta may be emitted. Allowed transitions between states are governed by certain selection rules to be discussed shortly.

Experimental confirmation of the existence of a series of different

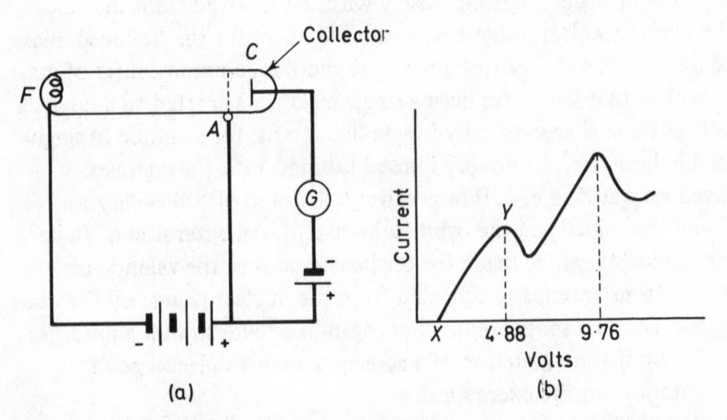

Figure 1.2 (a) Circuit for excitation potential measurement, (b) excitation potentials in mercury vapour

energy states was demonstrated by Franck and Hertz in 1913. If electrons from a filament F are accelerated by an anode grid A (*Figure 1.2(a)*) and bombard atoms of some gas, say mercury vapour, at a pressure such that the mean free path is considerably less than the anode spacing (in order to increase the opportunities for collision) then, if a collector C close to A is biased about 0.5 V negative with respect to A, it will only allow electrons with energies exceeding 0.5 V to flow in the galvanometer circuit G. *Figure 1.2(b)* shows that the current increases continuously until the voltage on A reaches 4.88 V, soon after this it falls and rises again until at 9.76 V a second peak occurs. The range X to Y results from elastic collisions; at or just above Y, inelastic collisions occur between electrodes F and A, i.e. energy is absorbed from the bombarding free electrons and given to an atomic electron. This is therefore raised to an excited state 4.88 V above its previous level, and is indicated by the fall in galvanometer current since the 'free' electron has not sufficient energy left to surmount the 0.5 V potential barrier to get into the galvanometer circuit. The atom in falling from its excited state emits a photon of radiation of energy 4.88 eV, corresponding to a frequency

$$\nu = \frac{4.88}{6.62} \times \frac{1.6}{10^{19}} \times 10^{34}$$

and hence a wavelength

$$\lambda = \frac{c}{\nu} = \frac{3 \times 10^{8} \times 6.62 \times 10^{19}}{4.88 \times 1.6 \times 10^{34}} \times 10^{10}\ \text{Å}$$

$$= 2537\ \text{Å}$$

which has been detected in the vacuum ultra-violet.

The second peak occurs at 9.76 V, twice the resonance potential of the first peak, and is the result of two inelastic collisions. Further transfer of energy in this way gives the whole series of energy states above the ground state, i.e. reveals the existence of a discrete energy level system which differs from atom to atom.

These simple ideas form the basis of our understanding of the origin of spectral lines, terms and series in optical spectra. Fine structural details of spectral lines and their 'splitting' by magnetic fields (Zeeman effect) can be explained by refinement of the Bohr theory to take account of relativistic effects, and nuclear and electron spin.

1.3 The Quantum Numbers

We have so far met n, the principal quantum number, which on the Bohr theory (and for all hydrogen-like atoms) determines the energy in the 'orbit' and the spacing between the point charge nucleus and the electron. There are, however, three additional quantum numbers; the set of four are important because they act as an index of the energy states possessed by electrons. An extension of the Bohr theory by Sommerfeld in 1916 included elliptical orbits (as in the solar system). He replaced n by the sum of the orbital momentum, k, and the radial momentum, n_r, quantum numbers, that is, $n = (k + n_r)$ and showed that this led to the existence of more than one allowed state for a given value of n without affecting the value of the energy.

Later developments have shown the shortcomings of the Bohr–Sommerfeld theory, for example, the physically clear-cut orbits cannot be reconciled with the Heisenberg Uncertainty Principle, which postulates orbital momentum states when $n = 1$; furthermore, it cannot deal adequately with many-electron atoms. With the introduction of wave mechanics over the following 10 years and the proposed concept of electron spin by Uhlenbeck and Goudsmid in 1925, it has been possible to build up a concise theory of spectra of one-electron atoms based on the vector model of the atom. This accounts for, and also predicts, the direction of changes which can occur between states, involving the emission and absorption of radiation.

Conceptually, it is a convenience to think in terms of the orbital motion, and the development has proceeded in the following way. In virtue of the orbital motion, the valence electrons possess angular momentum l which is quantised, that is, takes values 0, 1, 2 . . . $(n - 1)$ units of $h/2\pi$. Thus we may say that this gives rise to shells of electrons, in which, for each value of n, there are n angular momentum energy states arising from the n values of l; as each electron has its own unique wave function, their overlap (and relativistic effects) produces a small range of energy over the range of l values. We name the separate shells K, L, M, N, . . ., corresponding to $n = 1, 2, 3, 4$ . . . and the electrons as s, p, d, f . . . corresponding to $l = 0, 1, 2, 3$ . . . In addition, the electron has another quantum number because of its angular spin moment, s, which takes only two values $\pm\frac{1}{2}$ units of $h/2\pi$, with the result that for any one electron there is a coupling of orbital and spin vectors which yields a total angular momentum vector $j = l \pm \frac{1}{2}$, depending on the orientation of the spin with respect to the orbital motion.

For quantitative calculations we must use the quantum mechanical values for *l*, *s* and *j* which work out as $\sqrt{[l(l+1)]}$; $\sqrt{[s(s+1)]}$; $\sqrt{[j(j+1)]}$ respectively. Both *l* and *s* precess round *j* in random directions in the atom (*Figure 1.3*) and the various states of the combinations are not

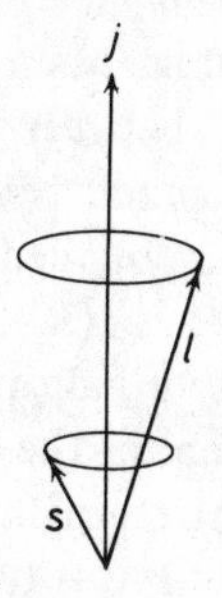

Figure 1.3 Precession of l and s vectors

resolved. The mere fact that we are dealing with a revolving and spinning charge means that we have a magnetic moment associated with the orbital and spin motions. Because magnetic interaction is possible between these internal fields, both vectors can couple to an external magnetic field which furnishes an axis, i.e. a fixed direction in space. The spin vector aligns itself parallel or anti-parallel to the field, while the *l* vector precesses round this axis in a limited number of orientations in space with respect to it: this means essentially that we have a limited number of quantised components of *l* along this magnetic axis (the perpendicular components neutralise each other). These range from +*l*

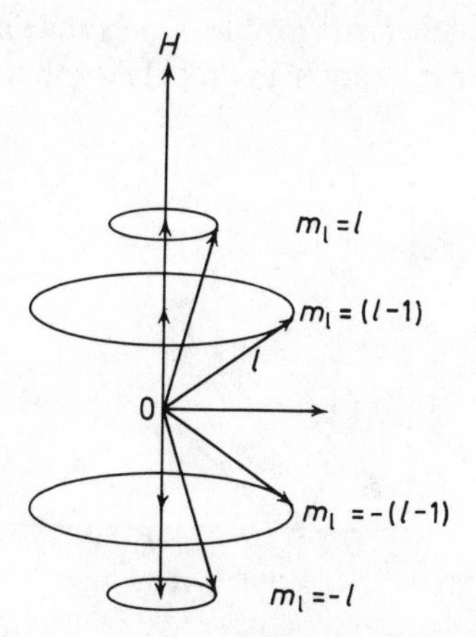

Figure 1.4 Space quantisation of states

through zero to $-l$ in integer steps, and are designated by the quantum number m_l; there are accordingly $(2l + 1)$ different space quantised states (*Figure 1.4*), and the electrons are called 1s for $n = 1, l = 0$; 2s for $n = 2, l = 0$; 2p for $n = 2, l = 1$; 3d for $n = 3, l = 2$, and so on.

When dealing with energy states in the shells of many-electron atoms of low atomic number, the pattern of coupling is that based on the Russell–Saunders scheme. Here, magnetic interaction between the l vectors exceeds that between l and s and we have a separate coupling for each, giving resultants $L = \Sigma l_i$ and $S = \Sigma s_i$; the compounded total angular momentum J takes values $(L + S), (L + S - 1) \ldots (L - S)$. When $L \geqslant S$ there are $(2S + 1)$ values for J. This multiplicity is indicated by placing a superscript to the left of the symbol representing the L value, the actual value of J is placed as a subscript at the bottom right-hand corner, for example, $^3p_1$ indicates a state in which $L = 1, S = 1, J = 1$. In heavier atoms where the spin–orbit coupling exceeds that between l vectors, we combine the j values for individual electrons in the shell to give a resultant J (jj coupling); as before, the $(2J + 1)$ states are resolved by a magnetic field to provide components M_J. This splitting of states in a magnetic field is the basis of the Zeeman effect.

We can form a mental image of the various states by examining the results of the solutions of the Schroedinger wave equation, remembering that the probability per unit volume of finding the electron in any particular 'state' is described by referring to the wave amplitude in the form $|\psi^2|$. This yields a picture of deformable electron cloud charge distributions of various shapes, each of which corresponds to a particular state. A schematic profile of a 2s state of hydrogen is shown in *Figure 1.5*; there is a very high probability that the electron will be

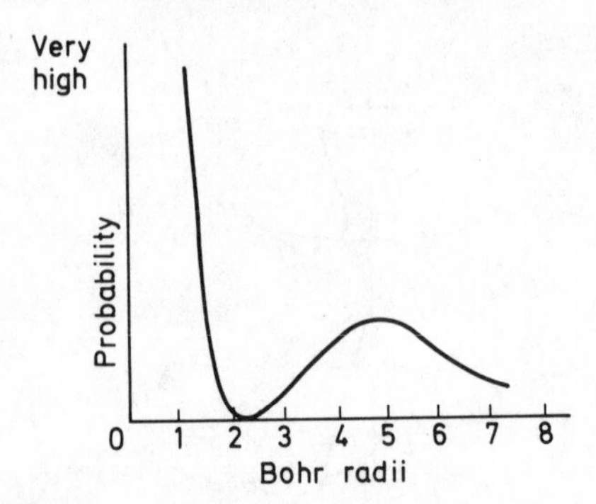

Figure 1.5 Probability of finding an electron in state (n = 2, l = 0) at various distances from the nucleus

found within a distance of one Bohr radius, but zero probability at twice this distance and a finite probability at five times this spacing, very different indeed from the Bohr predictions.

The quantum mechanical description automatically accounts for discrete energy states and points to a representation of the dynamical state of each electron in the atom in terms of four quantum numbers: n, l, m_l, m_s, which, according to the Pauli principle, are unique for each electron. Application of this principle leads to the possibility of determining the number of electrons in each nl state, that is, to the number of equivalent electrons in each shell. Further, it gives information on the extent to which the shells of different species of atoms in the Periodic Table are filled. Recalling that l takes values $0, 1 \ldots (n-1)$; $s = \pm\frac{1}{2}; j = l \pm s$, and using the spectroscopic notation $l = 0$ for s states, $l = 1$ for p states, $l = 2$ for d states, $l = 3$ for f states, and so on, we can build up the system of electron states starting with the hydrogen atom which has a single electron in the K ($n = 1$) shell closest to the proton, and a corresponding l value of zero, so that $j = +\frac{1}{2}$; these are the 1s electron states. The next element, He, has two 1s electrons in the K shell, for each of which $l = 0$, so that j has the two values $+\frac{1}{2}, -\frac{1}{2}$. A third electron would have to go into the next, L shell, as a 2s electron, to avoid violating the principle; it would have $n = 2, l = 0, j = \frac{1}{2}$, and the electron configuration would be that of Li. A fourth electron could be accommodated in this shell as a 2s of opposite spin, to yield the element Be, whose electron configuration, 2(1s) + 2(2s) is usually written with a superscript to denote the number of electrons, as $1s^2\ 2s^2$. No further s electrons can be added to this portion of the L shell; however, we may add a p electron with $l = 1$ to start up the p sub-shell of the main $n = 2$, L shell. Since the quantum number m_l can take values $+1, 0, -1$ and there are two spin directions, we could have six p electrons in this sub-shell and it is successively filled, one electron at a time, to yield the elements B, C, N, O, F, Ne. In general, a shell when filled, can hold $2n^2$ electrons, distributed among its various sub-shells, each of which can hold a maximum of $2(2l + 1)$ electrons.

As we proceed up the Periodic Table, therefore, we add a single electron to the sub-shell for each unit increase in atomic number (to keep the atom in a neutral state), until the shells hold their full complement of electrons and we should expect and, in fact do find, that the elements built up in this way show periodic properties determined by their shell structure, for example, Li, Na, K, . . . have a single valence electron which characterises their chemical and physical behaviour. It should be mentioned that there are several exceptions to this regular filling up of shells; for example, if a 4s level of one atom lies lower in energy than a 3d level of its neighbour, electrons will enter the former in preference to vacancies in the latter (*Figure 1.6*); the stability of the

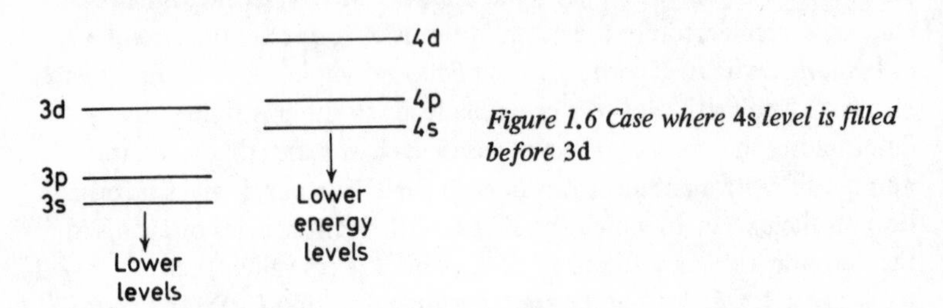

Figure 1.6 Case where 4s *level is filled before* 3d

system depends on the minimum energy principle. Some configurations based on the shell model for isolated atoms are given in *Table 1.1.* Each completed shell will have zero orbital angular momentum and zero spin, i.e. $L = 0, S = 0$; if this should happen to be the outermost shell of an isolated atom, the electron charge distribution will be spherically symmetrical, and we would expect it to show little or no tendency to combine with other atoms unless drastic measures are taken to alter the spin state of one or more electrons.

Where there are missing electrons in the outer shell of a neutral atom, we have the possibility of providing 'locating sites' for the easily detached, loosely held electrons of a different species of neutral atom; a familiar example is Cl, $1s^2$, $2s^2$, $2p^6$, $3s^2$, $3p^5$ and Na $1s^2$, $2s^2$, $2p^6$, $3s^1$ which, on coming together become negatively and positively ionised respectively, since the Na 3s goes to fill the 3p sub-shell of Cl; the result is an ionic crystal, $Na^+ Cl^-$.

1.4 The Selection Rules

The two main features of the Bohr theory are:

1. The introduction of the concept of energy levels in which the electrons are in stationary non-radiating states.
2. The postulate that transitions between states result in the emission or absorption of quanta in discrete amounts according to $E_i - E_f = h\nu$.

Table 1.1 SOME CONFIGURATIONS BASED ON THE SHELL MODEL FOR ISOLATED ATOMS

Element	*Atomic number*	K *shell* $n = 1$	L *shell* $n = 2$		M *shell* $n = 3$			N *shell* $n = 4$			
		$l = 0$ 1s *electrons*	$l = 0$ 2s *electrons*	$l = 1$ 2p *electrons*	$l = 0$ 3s *electrons*	$l = 1$ 3p *electrons*	$l = 2$ 3d *electrons*	$l = 0$ 4s *electrons*	$l = 1$ 4p *electrons*	$l = 2$ 4d *electrons*	$l = 3$ 4f *electrons*
Li	3	2	1								
C	6	2	2	2							
Na	11	2	2	6	1						
Mg	12	2	2	6	2						
Al	13	2	2	6	2	1					
Si	14	2	2	6	2	2					
Fe	26	2	2	6	2	6	6	2			
Cu	29	2	2	6	2	6	10	1			
Ge	32	2	2	6	2	6	10	2	2		
Se	34	2	2	6	2	6	10	2	4		

In this way it is possible to explain rather convincingly the emission lines of the spectrum of atomic hydrogen and to lay down a general pattern for the interpretation of optical spectra. The arbitrary nature of the Bohr assumptions has, of course, given way to the approach from wave mechanics; this is based on an appreciation of the wave-like properties of the electron and has confirmed the presence of an energy level scheme.

The question of the allowed and forbidden transitions between levels was answered by the quantum mechanical analysis, and resulted in the formulation of a set of selection rules governing the possible transitions from excited to normal states of the atomic system. These furnish criteria for a deeper exploration of line spectra; the main conclusions may be concisely stated as follows:

1. There is no restriction on n, the principal quantum number.
2. Allowed transitions result when the change in total orbital angular momentum quantum number, as between states, $\Delta l = \pm 1$; $\Delta m_l = 0, \pm 1$. This rule can be understood in the light of the law of conservation of angular momentum. An emitted or absorbed photon carries one unit of angular momentum with it; furthermore, as we are dealing essentially with electric dipole transitions in atomic spectra, allowed transitions can occur only between states of opposite parity. This notion of parity is derived from the symmetry relations of the wave function; a state has odd parity when its total l is an odd number, and even parity when it is an even number, hence the change in l brought about by a permitted transition is ±1. From these considerations, it will be seen that permissible transitions would arise between s($l = 0$) and p($l = 1$) states; between p($l = 1$) and d($l = 2$) states, but not between s states or between s and d states.
3. The change in the total angular momentum quantum number in allowed transitions, $\Delta J = 0, \pm 1$; $\Delta M_J = 0, \pm 1$, with the proviso that $J = 0$ to $J = 0$ is forbidden. This rule would therefore allow transitions between levels $P_{\frac{1}{2}}$ and $S_{\frac{1}{2}}$, also between the pair $P_{\frac{3}{2}}$ and $S_{\frac{1}{2}}$ since the lower right-hand subscript represents the j value for the state.

Although these criteria were developed to explain optically excited spectra, they also apply without modification to transitions in the deep-seated X-ray levels where the atoms are already in ionised states. We should, therefore, expect the difference between the two types of spectra to be largely one of scale; i.e. of the magnitude of the energies involved and hence of the wavelength range covered. Whereas in optical spectra energy values are of the order of the ionisation potential, say up to 10 V and wavelengths are in the range 3000–7000 Å, the usual 'hard' X-ray region involves energies in the order of tens of thousands of volts. In between these extremes lies the soft X-ray region, from say 50–500 Å, with which this book is mainly concerned. The broad principles of interpretation of X-ray spectra is thus based on the rules which apply in the optical region and, indeed, our ideas of electron configuration and structure of the valence band of solids, as unfolded by skilled researchers from the time of Moseley to Siegbahn in our own day, depend on the knowledge that these principles remain unchanged throughout the whole spectral range.

1.5 The X-ray Region

The transitions between energy states so far considered have been concerned with those in isolated atoms far apart from their neighbours, which have been excited in a discharge tube of some kind or by thermal means. Each spectral line registered by the spectrograph represents the resultant radiation emitted from a very large number of atoms, all in the same state, which undergo exactly the same transition in reaching their final state. As the levels involved among such valence states are discrete and sharp, it might be expected that the range and intensity of emitted radiation would reflect the density of states among the valence electrons. This is not so because the end points of the transitions for each emitted photon can vary, and it would be exceedingly difficult to sort out, in any consistent fashion, the number of electrons in each state which participate in allowed transitions. Further, the valence electron distribution and energy states in individual atoms change when the atoms come together to form the solid state; this follows from the superposition of the wave functions of the individual electrons giving a quasi-continuous range of energies forming a band. In order to get a faithful picture of this range it is essential that the end point of the transitions for all electrons involved should be the same, and that it should be a sharp lower level; this condition is fulfilled in the soft X-ray region.

Before discussing details of the transitions in the solid, it is of interest to get a broad picture of the X-ray region and understand how ideas developed by Bohr for the optical region have been applied successfully to explain changes in the deep-seated X-ray levels. It has been pointed out that the energies of ionisation for electrons much closer to the nucleus greatly exceed the values for optical ionisation, so that it is permissible to consider the state corresponding to optical ionisation as the zero level. Atoms in a solid which have one of their innermost K electrons ($n = 1$) removed entirely from the solid by the direct hit of bombarding electrons of energy E_K with respect to the solid, will be put into a K ionised state which will possess an energy E_K above the zero reference level. If electrons from the L shell ($n = 2$) fall into these vacancies, then the solid will emit quanta of radiation equivalent to the differences in energy between the K ionised and L ionised states, since at the end of such transitions, the atoms would find themselves in L ionised states of lower energy. Clearly, the possibility exists for electrons from shells still further away from the nucleus, for example, M($n = 3$), N($n = 4$), to furnish electrons to replace those knocked out of the K shell. This is what actually happens in the X-ray region, and the ensuing transitions produce the K series of X-ray spectral lines whose frequencies and hence wavelengths, depend only on the energy differences between states in the shells; the larger this difference, the shorter the wavelength of the observed radiation. As the atomic number Z and hence the nuclear charge increases, the binding of the inner electrons becomes stronger and this has the effect of increasing the energy difference between the shells compared with that which exists for low Z values. Similarly, the L series of lines is produced by transitions which end in the L shell.

In dealing with the level system in solids we need to modify the Bohr formula for quantitative work, since this was established for hydrogen-like atoms. The field 'seen' by a single valence electron in the solid will depend on its position; it comes under the combined influence of the nuclear charge and that of all the other electrons, and as the latter screen the nuclear field, we must introduce a slight modification of the Bohr theory to allow for this shielding effect. This is done by replacing the atomic number Z by a smaller number, $(Z - \sigma)$, where σ, the screening factor, will vary with the position of the electron (approximately unity for a missing K electron). In the transition from the L($n = 2$) to the K($n = 1$) shell which results in a line of the K series, the wave number is given by

$$\bar{\nu} = (Z - \sigma)^2 R\left(\frac{1}{1^2} - \frac{1}{2^2}\right) = \tfrac{3}{4}R(Z - \sigma)^2$$

and if the atomic number Z is varied, we should get a straight line plot between the square root of the observed frequency ν and Z. From this the value of R can be derived and checked; alternatively, knowing R, we can derive Z.

Experiments based on these deductions from the Bohr model were carried out by Moseley in 1916, who recorded the X-ray lines in the K, L, M . . . series of a large number of elements. He showed conclusively that the square root of the frequency for any line in each series increased with Z and established the relation $\nu^{\frac{1}{2}} = C(Z - \sigma)$ experimentally. Both C and σ are constants for any given series, but have different values for each series, since the screening effect of the extra-nuclear electrons increases with distance from the nucleus.

It is a simple matter to evaluate the energy differences between shells when the emitted wavelength λ(Å) has been recorded. The Bohr quantum condition gives

$$E = \frac{hc}{\lambda} = \frac{6.62 \times 10^{-34} \times 3 \times 10^{8} \times 10^{19} \times 10^{10}}{1.6\lambda} \doteqdot \frac{12\,400}{\lambda}$$

As an example, the K_{α} line in the soft X-ray spectrum of carbon results from a transition involving the transfer of a p electron of the L shell to an s level in the K ionised shell, thus leaving a vacancy or hole in the L shell. The wavelength emitted is about 45 Å, so that the energy difference between the K ionised and L ionised shells is about 270 V. Above this threshold energy, the number of transitions taking place will depend on the number of primary electrons which impinge on the carbon target.

Returning now to X-ray transitions in the solid, it is important to remember the necessity of having a sharp lower level for the end point of the transition; this only occurs in the soft X-ray region. In the case of carbon, for example, the terminal K level is only about 0.08 V wide, whereas the L level which contains the valence electrons extends over some tens of volts and contributes p electrons to the narrow K band; the result is a range of quanta contained within a band of K radiation which must reflect the energy spread of their source, that is, of the L band. As our experimental procedures provide a measurement of the distribution of the emitted intensity from the carbon (or other targets), we can obtain information about the width and the variation of the density of p states within the L band.

The question of allowed and forbidden types of transitions in X-ray spectra is again answered by the quantum mechanical analyses. Briefly, the same selection rules that apply to optical spectra hold good, i.e.

allowable transitions must have $\Delta l = \pm 1$, $\Delta j = \pm 1$ or 0, with no restriction on *n*. When applied in this way, they account satisfactorily for the large majority of transitions which are associated with the emission of X-rays. A scheme of energy levels covering the K, L and M bands is shown in *Figure 1.7*, in which the quantum numbers, *n, l, j* are given for each

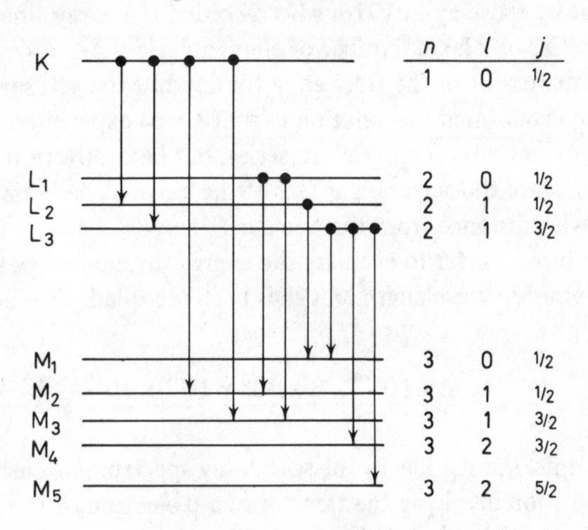

Figure 1.7 K, L *and* M *transitions*

level. It will be observed that the L shell has three levels, L_1, L_2, L_3, corresponding to the three values of *j* for the electrons. Similarly, the M shell consists of five separate levels, M_1, M_2, M_3, M_4, M_5, associated with the five values of *j*. The K band would result from p state transitions in L and M to the s states in K and therefore would begin on $L_2(\Delta l = 1, \Delta j = 0)$, continue on $L_3(\Delta l = 1, \Delta j = 1)$ and then proceed via $M_2(\Delta l = 1, \Delta j = 0)$ and $M_3(\Delta l = 1, \Delta j = 1)$. Where bands overlap and there are mixtures of s, p or d states, only those transitions occur which satisfy the selection rules; some evidence of cases of this kind will be presented later.

The information on band structure afforded by study of emission spectra is exclusively confined to knowledge about the population of states in filled levels; to get information about the unfilled, i.e. empty levels, requires investigations into absorption spectra. In absorption, the quantum of radiation incident on the solid must be energetic enough to dislodge an electron from one of the filled shells and place it in a vacant level above the topmost filled band; the resulting absorption spectrum arising from a multitude of such events supplies information about the

disposition, density and width of the empty level structure, but interpretation is not straightforward. The presence of absorption edges and their energy can be demonstrated by simple experimental techniques in which we measure the absorption coefficient per unit mass of material, μ_m, by irradiating thin foils of it at different wavelengths, i.e. at

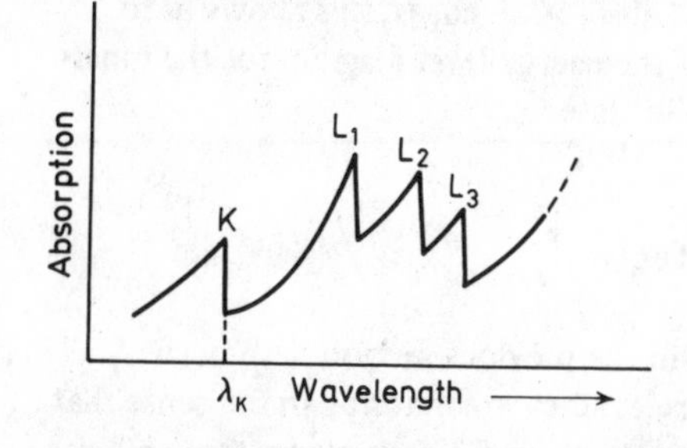

Figure 1.8 K *and* L *absorption edges*

different quantum energies. The coefficient is determined from the incident and transmitted intensities I_0 and I_t respectively, using the relation

$$I_t = I_0 \exp(-\mu_m d)$$

where d is the thickness of the foil expressed as mass per unit area. A typical curve (*Figure 1.8*) illustrates the form of variation of μ_m with incident quantum energy. The absorption increases as the cube of the wavelength until the K band edge is reached, and then falls steeply; the value at the 'edge' defines the binding energy of the electrons in the K shell. As the wavelength is further increased, the cubic law is again followed until the L_1 edge is reached, at which there is a repetition of the abrupt fall in μ_m; as before, a precise measurement of the wavelength at this 'edge' serves to determine the energy of the L level. The same form of the variation occurs for the L_2 and L_3 edges. In similar fashion, it is possible to confirm the existence of the five M edges defining the M levels.

When the incident quantum has an energy in excess of that necessary to ionise in any particular shell, the electron released will carry away an energy equal to the difference between the incident energy and the ionisation energy of that shell. Its energy can be determined from an analysis of its path in a uniform magnetic field; such observations have opened up the new field of electron spectroscopy. A magnetic analysis of the photoelectrons liberated from the solid when it interacts with quanta of known energy content, can give information of the binding energy and density of states comprised by the whole system of levels. Further comment will be made in the last chapter on this method of exploring valence states in solids. Precision is given to our inferences about the structure over the whole system of levels if, in addition, we combine with the magnetic data the knowledge obtained from an examination of the emission and absorption edges; this allows us to draw a fairly accurate picture of the energy level diagram for the inner atomic levels of atoms in the solid state.

1.6 Satellites and Auger Effect

The X-ray transitions discussed in the previous section have been accounted for by the selection rules; they are 'natural' in the sense that the energy changes which give rise to them follow a set pattern and we may term them 'diagram' lines. However, in an actual spectrogram, in simple cases, invariably on the short wavelength side, one sometimes finds weak 'foreign', additional lines in the vicinity of the allowed ones, which have been named 'satellites'. The same type of irregularity occurs in the soft X-ray spectra of some metals, again on the high energy side of the maximum distribution of states, with an intensity of only about 1% of that at the peak.

The existence of these satellites must, therefore, be the result of transitions involving higher energy differences than those which give rise to the 'natural' lines, and it is not obvious how they can arise from energy differences in the diagram lines. Considerations based on the effects of multiple ionisation indicate that their presence results from transitions taking place between multiply-ionised states. For example, an atom which has suffered double ionisation, that is, has electrons missing from say, K and L shells, would be left in an energy state KL; if now an electron from the L shell should fill the vacancy in the K shell, the atom would be left in the doubly ionised state, LL, in which two electrons are missing from the L shell. The transition KL to LL involves a larger energy difference than the normal K to L change, and therefore radiation of shorter wavelength is emitted; this gives rise to a satellite line and it is 'weak' because these events are few in number.

Finally, mention should be made of the Auger effect, which has been called internal conversion. This was first discovered in cloud chamber tracks following the interaction of X-rays with matter. It was found that when the X-ray excitation took place, the number of emitted quanta was less than the number of excited atoms produced. This meant that a proportion of the excited atoms in falling to lower energy states got rid of their excitation energy without radiating quanta. For example, an

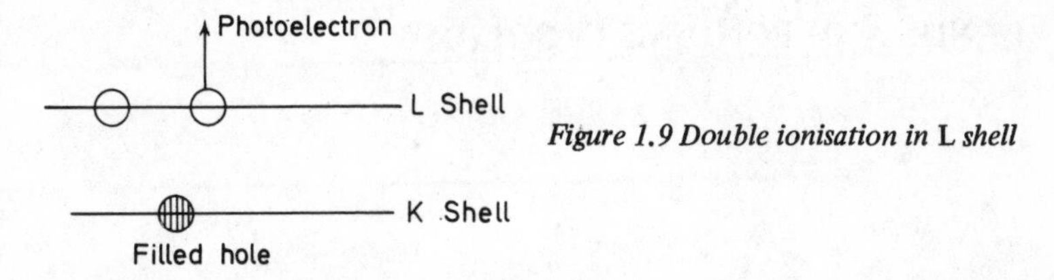

Figure 1.9 Double ionisation in L *shell*

atom ionised in the K shell (energy K) can have its vacancy or hole filled by an electron from the L shell; in returning to the state of lower energy it may release another electron from the L shell with a kinetic energy, $K - LL$ (*Figure 1.9*). These changes without definite radiation of energy are called Auger transitions, and there is no question here of an emitted quantum which acts on the atomic level system to give rise to photo-electric emission. Again, it is possible, using magnetic analysis to obtain information on the LL state. Further reference to the Auger effect mechanism will be cited later when dealing with the interpretation of the band forms of solid state spectra.

Chapter Two

BANDS AND BONDING IN SOLIDS

2.1 Introduction

We have seen that the energy levels occupied by electrons in isolated atoms are governed by quantum conditions, that is, the electrons are restricted to certain allowed states which are unique for each electron; these can be derived from the solution of the Schroedinger equation which, in principle at least, can be solved if the combined potential field arising from the nuclear charge and the extra-nuclear electrons is known. The effect of including the interaction effects between the outer valence electrons is to make the solution more difficult; however, perturbation methods have been developed to deal with these cases. In the simple example of an individual atom with two electrons in different states, that is, with different wave functions, and with a repulsive interaction energy e^2/d, where d is their distance apart, then, as the electrons can change places and are indistinguishable, we must introduce a new wave function and an energy of exchange to describe the combined state. Application of the Pauli exclusion principle, together with considerations based on the configurations allowed by the vector model of the atom, lead to the result that the two outer electrons must have their spins parallel in order that the system as a whole may have its lowest energy, and this is the most stable configuration. In fact, the various 'ground' states of atoms can be decided from the criteria above, which involve the spin orientation of their outer shell electrons, for example, the two p electrons in the ground state of the carbon atom have parallel spins, and the state of lowest energy would therefore be the triplet ^3P state.

In order to understand how the energies of the electrons in the various atomic levels are affected when atoms come together to form a solid it is, as a preliminary step, instructive to examine what happens to the energy states of two hydrogen atoms when they come together to form a molecule. An understanding of this simple system of similar atoms will help us to transfer concepts in a general way to the vast conglomerate of atoms that go to make up a solid body.

2.2 The Hydrogen Molecule

When the two atoms of hydrogen are brought closer and closer to each other the outer valence electrons are the first to react with each other, followed by attractive interaction between each electron and the 'other' nucleus, and repulsive effects between nuclei. It turns out that the overall interaction energy increases when the spins of the two electrons are parallel; this results in a weaker bond between the atoms, and therefore tends to prevent formation of the molecule. On the other hand, when the spins are anti-parallel, the interaction energy is decreased, the binding is increased and the atoms attract each other to form a stable molecule; the equilibrium separation is determined by the condition that the energy of the system as a whole must be a minimum.

The behaviour of the wave systems of the electrons can be understood by considering the way in which their separate wave functions interact as the two atoms approach each other. This is illustrated schematically in *Figure 2.1*, where the upper diagram (a) shows the form of the potential curves for the isolated atoms, and how it changes for the combined system. The lower curves (b) represent the separate wave functions for the ground 1s states and the result of superimposing them in two ways: symmetrically, and anti-symmetrically. It will be seen that the combined wave function has high values everywhere between nuclei for the situation depicted in (a), whereas it diminishes rapidly to zero midway between nuclei for that depicted in (b); further, it has smaller values elsewhere when compared with the unperturbed wave functions. Since the probability of finding the electron in a given element of volume is determined by $|\psi|^2$, it follows that the combination which results in a symmetrical wave function must correspond with a high density of electrons between the nuclei and, consequently, with a lowering of the potential energy of the combined system.

The anti-symmetric superposition leads, on the other hand, to a small probability of finding the electrons between the nuclei so that the charge density there is low and the potential energy of the system has increased.

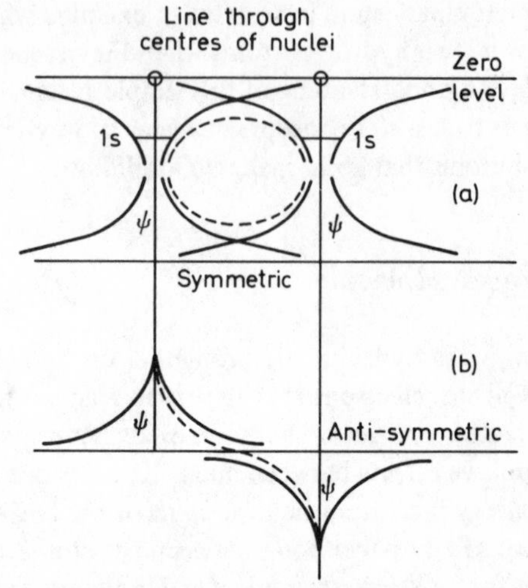

Figure 2.1 Superposition of wavefunctions (a) symmetric, (b) anti-symmetric

The most stable configuration must obviously be that associated with the combined symmetrical wave function and anti-parallel spin of the two electrons, since it embraces a reduction in the energy of the system and gives rise to the binding of the atoms in the molecule.

2.3 The Splitting and Overlap of Levels

The special feature of the way the wave functions combine should be stressed; there are two effects – a raising and a lowering of the energy E of the single 1s level of the separate atoms. In effect, the result of the superposition is to provide two distinct levels differing in energy – one below E, the other above E – which are created from the single energy level of the individual atoms by the mere act of allowing the wave functions of the electrons to interact with each other. This means that from the moment of overlap the single energy level separates into two levels whose energy difference increases with the degree of overlap, so that at the equilibrium distance these lowest and highest energy values define the limits of a band of energies.

It is possible to view all the states in the band as arising from changes in frequency of the waves following on the interaction between the electron waves; The situation depicted in *Figure 2.1(a)* indicates that the perturbed amplitude does not fall away as fast as the unperturbed one. This implies that the wavelength associated with the composite wave has increased; hence this way of superposing should result in a lowering of frequency and should therefore provide the lower limit of energy of the combined system: it accordingly represents the most stable configuration. The upper limit is derived by the kind of superposition represented in *Figure 2.1(b)* where the amplitude falls faster than that in the original wave. Here, the wavelength has decreased so that the frequency, and with it the energy, will have correspondingly increased; the result is a less stable system brought about by what may be called anti-bonding forces. The degree of lowering of energy when the atoms are at their actual equilibrium separation is determined by the balance between the attraction of the nuclei for the electron clouds and the repulsion between the clouds on the one hand, and between the nuclei involved on the other; the configuration finally taken up is that for which the system as a whole possesses minimum energy. It should be remarked in passing that this manner of bonding is characteristic of the covalent bond between certain types of atoms in solids.

This feature of the spreading out of the energy between defined limits can be said to be a characteristic of the effects of the interaction of the valence electrons; it is exhibited by atoms coming together to form the solid state. When the atoms involved are heavy, so that each contains several shells of electrons, the wave functions which represent the innermost electrons closest to the nuclei will scarcely have an opportunity of overlapping, since the physical space to which they extend is limited by their 'bound' condition, and we must therefore expect that the single character of the energy levels in the innermost shells, i.e. K and L, will still be maintained when such atoms take up their equilibrium positions in the solid. The valence electrons in the outer levels, being much less tightly bound will, however, react with each other, so that their wave functions representing corresponding quantum states can combine symmetrically and also anti-symmetrically to yield energy level limits lower and higher than their original common value – the difference in energy between the bottom and top of the 'split' level, that is, in the 'band', may amount to several electron volts. If it were possible to bring the nuclei physically closer and closer in stages, beyond the equilibrium value, then the inner levels would progressively begin to show the same type of broadening or splitting, commencing with the shells next to the valence band and working inwards towards the K shell. A schematic sketch of the broadening into bands is shown in *Figure 2.2.* The reduction in energy of the valence levels as exhibited by the lower branch V_l is predominantly responsible for the binding of the atoms; the upper branch V_u corresponds with the anti-bonding state, and should therefore represent the maximum energy, mainly in kinetic form, of the electrons in that state. There are thus two quantum states involved here, and each is capable of holding two electrons of opposite spin, making a total of four quantum states in all, which is the combined

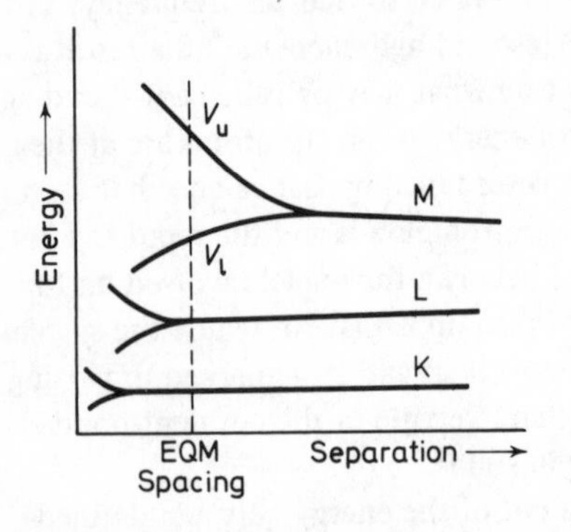

Figure 2.2 Splitting of levels

number in the individual atoms. It is correct to generalise and say that the total number of quantum states to be found in the entire quantum system is the sum of the separate number available in the component atoms; further, the energies corresponding to these quantum states must be distributed between the minimum and maximum energies which arise from the symmetric and anti-symmetric superposition of the wave functions of the valence electrons.

It should be realised that the confinement of the energies of interaction between certain well-defined limits in the solid is, in effect, a consequence of the Pauli exclusion principle, for the band of discrete energies between the limits is composed of alterations in energy in quanta; these arise from the range of frequencies which automatically 'fall out' in the process of combining the wave functions of the separate electrons. The limits of energy do not depend on the number of atoms involved; however, the more atoms, the larger the number of levels that have to be accommodated between the lower and upper limits of the band – with the proviso that not more than two electrons may occupy any one of them. Some idea of the numbers involved can be gleaned from the fact that 1 cc of an average metal would contain about 10^{23} atoms, so that some 10^{23} discrete energy levels exist between the lowest and highest limits in the band; each level can hold two electrons of opposite spin so that their total number will be about 2×10^{23}: this will represent the total number of quantum states in the band. If we take a band width of 10 V, the average spacing between levels in the band would be about 10^{-22} eV, and this separation extends throughout the whole of the crystal.

2.4 Insulator and Semi-conductor Bands

The systematic filling of energy levels with two valence electrons per level occurs in all cases. When this happens over the whole of the level system the band will obviously be full, for each atom contributes its quota of two valence electrons and there are as many levels as atoms in the lattice. Above this full valence band there will be an empty conduction band whose separation will depend on the type of material considered, for example, for insulators, some 5–10 V, for semi-conductors about 1 V. The energy gap in the former case cannot be bridged except by increasing the energy of the topmost electrons of the lower band by exceptional methods, and this means that it would show exceedingly poor conductivity under normal conditions – it would be classed as an insulator. A familiar example can be cited: the carbon atom has a full K shell and an electron configuration $1s^2\ 2s^2\ 2p^2$, and when uniting with others to form solid diamond has its p sub-shell filled by an addition of four electrons from its four closest neighbours, each contributing one, which resonates between the atoms to form the electron-sharing or 'covalent' band. In the solid, therefore, the L band is now full, and it should show insulating properties since the conduction band is situated some 6 V above the full band. This is a simple concept and explanation, though the real situation is somewhat more complicated. A mathematical analysis shows that the outer 2s and 2p bands begin to overlap at a lattice spacing exceeding the equilibrium spacing; the form of the calculated band structure is shown in *Figure 2.3* and can be interpreted as follows. As the isolated atoms approach each other the wave functions of the two electrons in the 2s and 2p levels begin to overlap, and the single state of each splits up into two as before. These have a 'crossover' at some lattice spacing exceeding the equilibrium value and yield a hybrid (2s, 2p) lower band which contains the maximum number of four available electrons, and an upper empty band separated from the lower by a forbidden energy gap of about 6 eV – a formidable one for electrons to cross under normal conditions – hence diamond shows insulating properties.

In semi-conductors such as Ge and Si the gap between the full, outer valence band and the next higher available empty band is of the order of 1 eV, and it is possible, even at ordinary temperatures, to transfer sufficient thermal energy from the lattice to electrons in the filled band to enable them to cross the gap and enter the empty band. Such promoted electrons would now be able to move about in the comparatively empty conduction band, so that these 'intrinsic' materials would therefore show a conductivity which increased with temperature; elements of this type are to be found in Group IV of the Periodic Table.

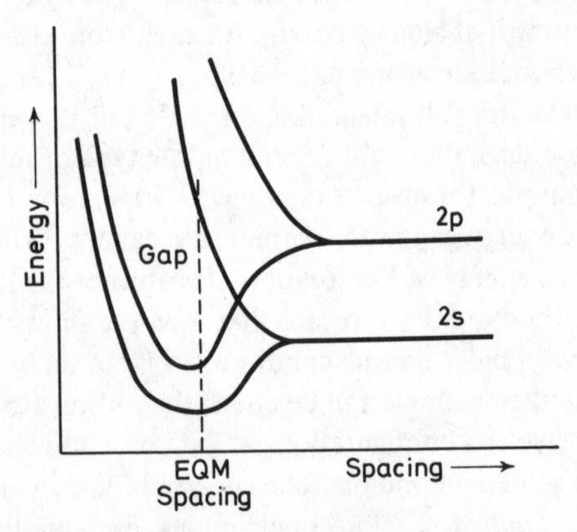

Figure 2.3 Energy band system for diamond

The conductivity of the 'intrinsic', that is, pure, semi-conductors can be greatly increased and controlled by substituting into their lattice certain types of 'impurity' atom in place of the host atoms. These may be either 'donors' like arsenic and phosphorus with five available electrons (where only four are necessary to fulfil the requirements of the bonding), each of which can donate a single electron for conduction purposes and yield n type material, or they may be 'acceptors' like boron and gallium with three valence electrons which can accept an electron from the full valence band, and in so doing create a hole in it, thus yielding p type material. These classes of impurities should therefore be considered as occupying or rather furnishing 'states' close to the bottom of the empty conduction band, or just above the top of the full valence band respectively. They introduce a collection of localised energy levels in their own immediate vicinity and reduce the gap width of the intrinsic material by factors of 10–20 or more. A schematic diagram of the band structure for the three cases is shown in *Figure 2.4.*

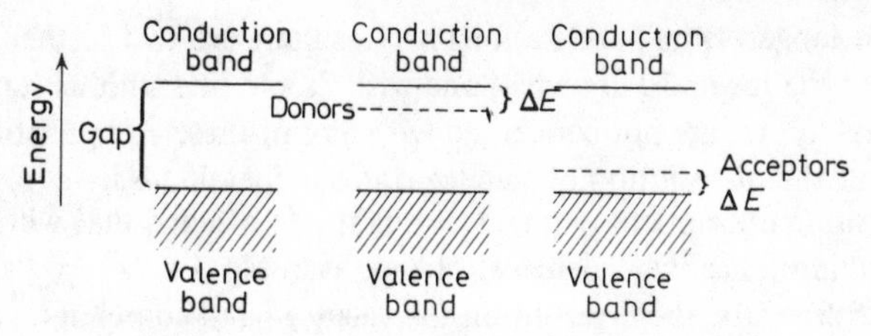

Figure 2.4 Band scheme for semi-conductors

The result of the doping is to produce an enhanced conductivity either through the availability of an increased number of electrons which enter the conduction band, or by transfer of an increasing number of holes in the lower filled band; the increase in each case depends on an exponential factor involving the new gap width and the temperature.

2.5 Metal Bands

One of the special features of the metal bands is that, unlike insulators, they are never full, since if there are not enough electrons to provide two per energy level per atom, then clearly the band of levels will only be partially filled, and in this case the element would show conduction in this partially filled 'conduction band'. A good example by way of illustration is the alkali metal Na, whose electron configuration in the isolated atom is $1s^2\ 2s^2\ 2p^6\ 3s^1$. Here the wave functions of the valence electrons in the 3s level of, say, N atoms of the solid overlap to form the valence band which can only be half filled, since each of the levels in it can hold two electrons of opposite spin, and only one per atom is available. This means that the first $N/2$ levels of the band will be filled, the remaining $N/2$ levels will be vacant and could therefore accept electrons from the lower half of the band, so that Na would behave as a good electrical conductor. In the band scheme of *Figure 2.5* the single 3s level splits into lower and upper branches corresponding with the 'bonding' and 'anti-bonding' orbitals. It will be noticed that the 3p splits earlier, i.e. when the separation between atoms is larger, and that the lower 3p branch overlaps the upper 3s and lies below it at the equilibrium spacing in the crystal. As the 3s band is only half filled, there will be little or no admixture of p states to interfere in the description of the valence range. At high excitation energies there are two further cross-overs of the lower 3d branch – one with 3s and one with 4s states; fortunately, we are not concerned with any of these complications in our study of the distribution of valence states in metallic Na.

Pursuing the earlier argument, it might be imagined that where the isolated atom has two valence electrons, as in Mg, $1s^2\ 2s^2\ 2p^6\ 3s^2$, then, in the solid state, these would fill the valence band completely, and we should therefore expect the material to be an insulator rather than a conductor. This does not happen on account of the overlapping of the

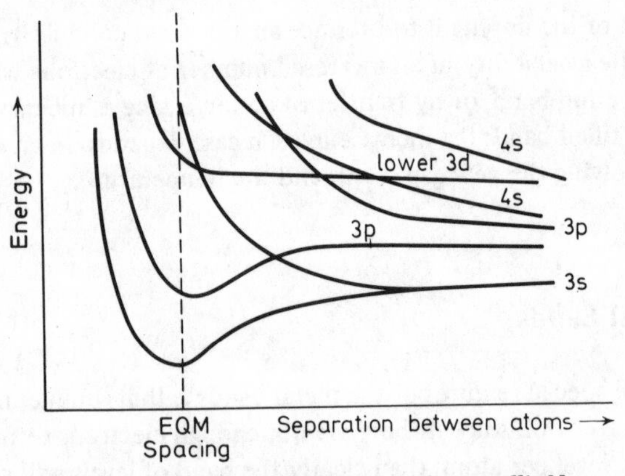

Figure 2.5 Band scheme for metallic Na

3s and 3p bands. In fact, this feature of overlapping bands for the higher quantum numbers, where the corresponding wave functions of the electrons populating the levels are physically significant over a range of the order of atomic dimensions, characterises the energy distribution of quantum states in the solid. As we have seen, there is a mixing of bands to the left of a 'crossover', and this sets a difficult problem – that of allotting the proportion of each set of states to be attributed to each band. In the case of magnesium the band structure reveals the fact that some of the 3s levels actually lie within the 3p band and above 3p states lower in the band; this combination of bands and mingling of states does, however, provide a range of empty levels in the upper portion of the mixed band into which electrons occupying lower levels can be promoted when their energy is raised. Here again we meet the property of electrical conductivity associated with partly filled bands, and we do, in fact, find that all elements subject to characteristic overlapping of this kind have conducting properties; their quality as conductors is determined by the extent of the overlap.

It is clearly of great interest to have experimental confirmation of the theoretical calculations of band structures. We wish to have sound evidence about such things as: (a) the width of the occupied portion of the band, derived from the range of energies to be found in going from its lowest level to the limit of the filled states; (b) the density of states; (c) the form and structure of the level system. Information on all these aspects of the band structure can be obtained from studies of solid state spectra as derived by the soft X-ray techniques to be described later. For example, work in this field on Na has shown that its band width agrees rather well with that calculated by treating its valence electron as a free electron. The width associated with the observed L spectra is the result of transitions taking place between the electrons in the lower half of the 3s band and the 2p sub-level of the L shell, since $p \rightarrow s$ transitions are allowed by the selection rules. As the L_3 level is single, the energy radiated must consist of sets of quanta, one for each electron making the transition, from each level to its end point at the single L_3 level; in this way the range of energies in the band can be recorded. We must emphasise the requirement that the final level in the transition should be sharply defined in energy if accuracy in band width determinations is to be achieved. This is best fulfilled in the soft X-ray region, for its breadth in a solid is mainly controlled by the effect of the lattice field and the radiation damping; it gives a measure of the resolution to be expected for the band. For example, a $p \rightarrow s$ transition in hydrogen-like atoms which involves a wavelength of 100 Å (the soft X-ray region) would correspond to a K breadth of 4×10^{-4} V, whereas a wavelength of 1 Å (hard X-ray) would have an end point level, some 4 V wide, and, in fact, of the order of width of the valence band being investigated.

We have seen that the 3p and higher levels are broadened in similar fashion to that of the 3s, and that the lower 3p may overlap the upper 3s at the equilibrium spacing of the atoms in the metal. As this is a characteristic feature of the bonding in all such solids, we should expect to find the level structure to be markedly altered when alloys are formed from different metals and also from non-metals. In fact, one of the current problems in this field of research centres round the extent of the modifications undergone by the band structure of each component as the valence and concentration are varied; the interpretation of results will obviously affect our ideas about the subsequent physical properties of the alloy.

2.6 Ionic Solids

Another class of solids whose structure can be fitted into the band picture we have built up has been named 'ionic' from the fact that the lattice sites are occupied by ions. These are formed by electron transfer from one component to the other when the crystal is formed; the resultant bond between the ions is predominantly an electrostatic attraction which leaves the system in the condition of minimum energy at the equilibrium spacing in the solid. Generally, combination between any element in Groups I or II with one in Groups VI or VII of the Periodic Table gives rise to the ionic bond; a common example is furnished by NaCl whose lattice consists of alternate Na and Cl ions which are formed when the single 3s electron of Na, $1s^2\,2s^2\,2p^6\,3s^1$, which is free, leaves its atom and is readily accepted by the Cl, $1s^2\,2s^2\,2p^6\,3s^2\,3p^5$, which needs it to complete its 3p sub-shell. The resultant alternation of positive and negative charges constitutes two interleaved face-centred cubic lattices of Na^+ and Cl^-; the electron configuration in each of these ions is now complete, and they thus have spherically symmetrical charge distributions. In such a lattice the interaction between the six nearest neighbours of opposite sign is that of a Coulomb attraction, while that between the 12 next nearest neighbours of the same sign is a repulsion, so that by summing over the whole lattice we arrive at a resultant attraction, as shown in branch *A* in *Figure 2.6.* When the

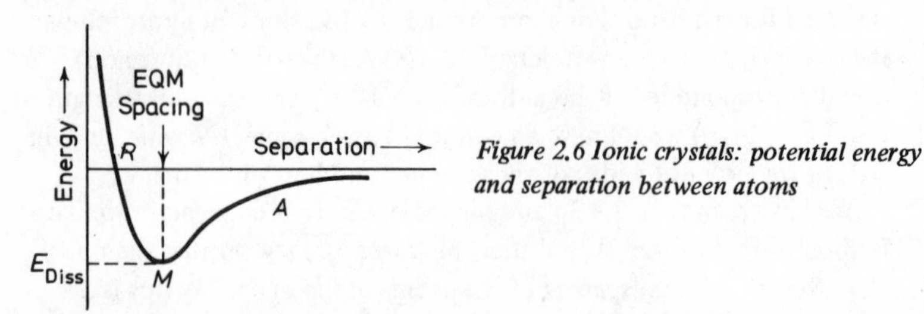

Figure 2.6 Ionic crystals: potential energy and separation between atoms

spacing decreases below its equilibrium value in the solid, the repulsion between the space charge clouds leads to a rapid increase in potential energy of the system, as shown by branch *R*; the minimum at *M* defines the equilibrium separation, while E_{Diss}, which corresponds with the state at *M*, represents the energy which has to be supplied to the crystal in order to dissociate it, that is, to break the ionic bond completely.

A schematic drawing of the band model for NaCl is shown in

Figure 2.7, where only the upper filled levels, 2s, 2p, 3s, 3p, are indicated. The crystal structure and the 'layout' of the internal fields are such that the wave functions of interacting electrons in all the bands must be concentrated in the immediate vicinity of the ions; there is consequently little or no build-up of electron density between them, and, since the gap between bands is quite large, these solids behave as insulators at room temperature, though conduction can take place at high temperature through the movement of ions in the lattice.

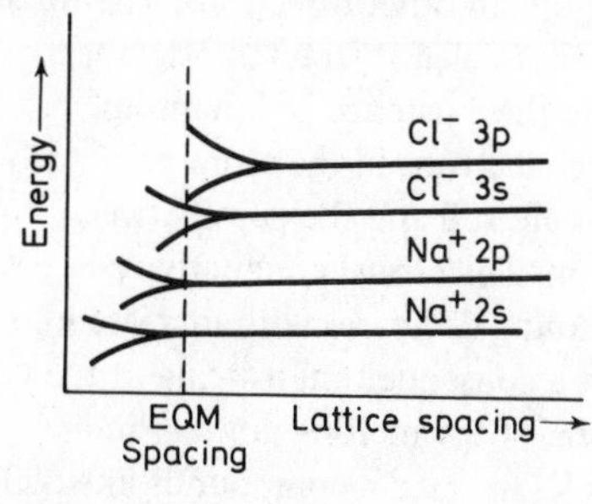

Figure 2.7 Band scheme for NaCl

It should be mentioned that the electrical properties of certain types of ionic solids, such as oxides, can be materially changed by incorporating specific types of neutral metal atom components into the lattice, either substitutionally or interstitially; the presence of neutral atoms of the metal component in either of these positions can introduce a localised impurity level just below the bottom of the empty conduction band, and therefore results in properties like that of an n type semiconductor. This sort of behaviour is also manifested when certain oxides are reduced so that a small fraction of the non-metallic constituent is removed and a non-stoichiometric compound produced. This possibility of changing the class of a solid in such a way as to confer special properties on it for certain applications, is a development of considerable technical importance.

2.7 Summary of Bonding in Solids

We have so far dealt with the main classes of solids whose band structure is of interest. These come under the headings of insulators, semiconductors, metals and ionic crystals, and it is in place in concluding this chapter to review the main types of binding which give these solids certain properties characterising their behaviour.

In insulators and intrinsic semi-conductors the electron pair or covalent bond is responsible for the cohesion of atom to atom. This is a strong bond linking atoms which need an additional electron from each of their neighbours to satisfy valence requirements. It projects in definite directions in space, for example, in Ge the bonds join atoms which form a tetrahedral structure comprising the four bonds, with two electrons resonating backwards and forwards between each pair of atoms; this results in a high electron concentration in the bond directions, and a negligibly small electron density in all other directions. The forces involved in holding the atoms together are electrostatic in origin, and the strong binding is accounted for by the existence of 'quantum mechanical exchange' effects between electrons in the atom pair. As there are few free electrons the bands are full and the gaps between bands relatively large; this results in small electrical conductivity in the pure material. The strength of the binding decreases with increase in atomic volume, and this brings about a consequent narrowing of the band which is reflected in the lowering of the melting point – in diamond 2500 K, in germanium 960 K. In certain compounds like SiO_2 the ionic bond is formed alongside the covalent and dilutes its strength; the same things happen in certain types of impurity semi-conductors and in organic compounds.

The ionic bond results mainly from direct Coulomb attraction between positively and negatively charged nearest neighbours. Its order of magnitude is roughly given by the ratio between the square of the charge and the equilibrium spacing, so that for a typical ionic solid like NaCl with an equilibrium spacing of 2.8 Å, the binding would be of the order

$$(e^2/4\pi\epsilon_0 d)\frac{\text{joules}}{\text{atom}} = \frac{(1.6 \times 10^{-19})^2 \times (10^{10}/2.8) \times (10^{19}/1.6)}{4 \times 3.14 \times 8.85 \times 10^{-12}} \doteqdot 5 \text{ eV}$$

per atom, and such a value is supported by data on the heat of dissociation of the salt. The fact that we are dealing here with closed shell structures possessed by each of the ions, means that the charge distribution in each is spherically symmetrical; under these conditions the interpenetration of the electron clouds of the outer electrons is limited and results in short-range repulsion forces which, at the equilibrium separation in the crystal, lead to an energy of interaction of only about one-tenth of that due to the attractive forces.

The binding in metals and alloys is brought about by the swarm of valence electrons circulating as 'free' agents between the fixed ion cores of the lattice; the availability of these 'free' carriers accounts for their high conductivity. The electron concentration is not confined to any favoured atom or group of atoms and, as the crystal contains both positive and negative charges and is therefore electrically neutral, the binding can be considered electrostatic in origin; however, unlike the covalent bond, the metal bond has no specific directional properties.

The calculation of energy levels in metals is complicated when the atom contains more than one valence electron because of the interaction effects. Considerable success has been achieved where the 'one electron approximation' has been applied, that is, the force field in which a single valence electron moves is taken as that which results from the combined effects of the nuclei and the 'smeared out' field of all the other electrons. Generally, increase in valency involves considerations of interactions between increased numbers of electrons; this, in turn, yields wider bands and tighter binding. As an example of the range of binding energies for common metals we mention Na, about 1 eV per atom, and Fe (where electrons are missing from the 3d sub-shell and covalent binding enhances the total binding) about 4 eV per atom.

Finally, to complete the classification of solids, it is of interest to mention the weak Van der Waals' forces operating between atoms in molecular crystals such as solid Ar. Here attractive forces originate through dipole–dipole interaction. This is brought about by the fluctuating dipole moments associated with the electron motion in closest neighbours among atoms and molecules. Binding energies are very low – of the order of 0.1 eV per atom.

Chapter Three

QUANTUM STATE DISTRIBUTIONS OF ELECTRONS

3.1 Introduction

The picture of the energy band model as outlined in the preceding chapter is based on the results of the wave-like interaction between the valence electrons of the individual atoms; it envisages a closely spaced gradation of levels extending over the whole range of energy, from the bottom to the top of the band. The question now arises as to how the energies of the multitude of stationary states, i.e. quantum states, are distributed over the band. To answer this question it is necessary to find the density of quantum states, i.e. the number present in unit volume of the solid, centred round each value of the energy of the state concerned. This can be achieved by introducing the concept of a 'phase space', that is, each quantum state electron has its three momentum and three spatial co-ordinates, p_x, p_y, p_z and x, y, z respectively, represented by a single point in a six-dimensional 'phase space' whose elemental volume is made up of the product of a volume in real space, together with a volume in momentum space, $dp_x \; dp_y \; dp_z \; dx \; dy \; dz$. This 'phase space' can be looked on as compounded of three separate two-dimensional sub-spaces, in each of which the elemental representative point has a momentum co-ordinate p_x, p_y or p_z, and a real space co-ordinate x, y or z. The wave associated with the electron motion in each of these sub-spaces is taken to be reflected when the electron reaches the boundaries of the crystal; such potential walls are separated by distance of, say, a, b, c in the x, y, z directions respectively. The superposition of incident and reflected waves produces the well-known pattern of standing waves; these have wavelengths λ related to both the spacing between the boundary walls and the momentum of the electron which describes its state: further, using the De Broglie relation $\lambda = h/mv$, the kinetic energy is given in terms of λ by $h^2/2m\lambda^2$.

3.2 The Phase Integral: Fermi Energy

The energy states of the electron defined by the wavelength λ can be easily calculated, for the first quantum state corresponds to a stationary wave with λ equal to twice the distance between boundaries, i.e. $\lambda = 2a$, if we take the wave moving in the x direction. The second state will have $\lambda = a$, the third, $\lambda = (2/3)a$ and so on, the nth quantum state will therefore have a corresponding wavelength $\lambda_x = 2a/n_x$, a momentum $h/\lambda_x = hn_x/2a = p_x$, and an energy $E_x = n_x^2h^2/(8ma^2)$; the change of momentum in going from one state to the next has a constant value $h/2a$, and can be represented in a diagram, such as *Figure 3.1,* where the points *P, Q, R, S* are plotted for the single sub-space, and represent the excursion of the constant momentum between the boundaries; integration over the closed path *PQRS* yields the phase integral $2p_xa = hn_x$.

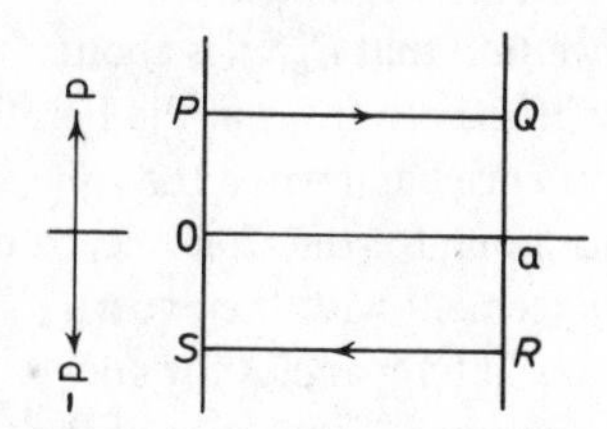

Figure 3.1 Two-dimensional phase integral

Extending this result to the other sub-spaces allows us to state the volume of the six-dimensional phase space, $n_xn_yn_zh^3$, which 'holds' the entire series of energy states in the band, where n_x, n_y, n_z are the components of the quantum number n defining the total energy of the state. The lowest state in the band has $n = 1$, and is associated with an element in momentum space of volume h^3/V, where V (= abc) is the three-dimensional volume in real space to which the electrons are confined; as n increases by unity from state to state, the volume in momentum space occupied by a single state remains constant at the value h^3/V.

The number of such elements, i.e. of quantum states, contained within a sphere whose radius in momentum space is p is clearly the volume of the sphere in momentum space divided by the volume of one quantum state

$$(4/3)\pi p^3/(h^3/V) \quad (3.1)$$

Since the surface of this sphere defines an energy $E = p^2/2m$ corresponding to the momentum p, the number of states whose energies lie within the limit E, per unit volume of real space, is

$$(4/3)\pi(2mE)^{3/2}/h^3 \quad (3.2)$$

As the Pauli exclusion principle allows two electrons of opposite spin to occupy each energy level, the number of electrons per unit volume of real space filling all levels up to the maximum energy E is

$$N = (8\pi/3h^3)(2mE_{max})^{3/2} \quad (3.3)$$

This maximum electron energy is, in metals, the Fermi energy, which thus has a value

$$E_F = (1/2m)(3h^3N/8\pi)^{2/3} \quad (3.4)$$

For Na, which has one valence electron per atom, N is of the order 10^{22} per cc and if we insert the constants, we find that E_{max} is about 3 eV. This value thus represents the energy of the electrons at the Fermi level, and therefore gives the width of the valence band when the bottom of the band is taken as the reference level; for the 'free' electrons of this theory we obtain reasonably good agreement with the experimental results for metals, but there are serious discrepancies for non-metals, for which the calculated values are too small. Some idea of the kind of agreement as between experimental and theoretically expected values for each case is presented in *Table 3.1*.

Table 3.1 COMPARISON OF CALCULATED FREE ELECTRON BANDWIDTHS WITH EXPERIMENTAL VALUES

	Metals					*Non-metals*			
	Li	Be	Na	Mg	Al	B	Diamond	Graphite	Si
Calculated band width, eV	4.8	14.6	3.2	7.3	11.9	25.2	29.5	21.9	12.7
Experimental band width, eV	4.2	14.7	3.0	7.4	12.7	31.0	33.0	33.0	18.2

3.3 The Density of Quantum States

The density of states in the band, i.e. the number of quantum states per unit volume which have energies between E and $(E + dE)$, usually signified by $N(E)dE$ is derived from the total number of states per unit volume:

$$N = (4\pi/3h^3)\,(2m)^{3/2}\,E^{3/2} \tag{3.5}$$

On differentiation, we get the density of states

$$N(E)dE = (2\pi/h^3)\,(2m)^{3/2}\,E^{1/2}\,dE \tag{3.6}$$

The actual number of electrons per unit energy range will, of course, be twice this value.

The way the density function $N(E)$ varies with the energy E of the state is shown in *Figure 3.2* when $T = 0$ K; the parabolic form rises to a

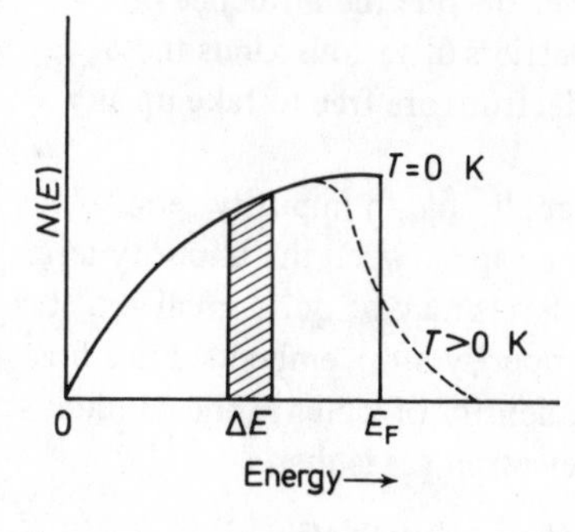

Figure 3.2 Fermi–Dirac distribution

'head' and then falls abruptly to cut the energy axis at a maximum value E_F representing the limit of the occupied states – the so-called Fermi energy: below E_F all energy levels are occupied, above it they are empty, so that the width of the valence band is given by E_F measured from a zero at the bottom of the band. It will be seen that the number of states per unit energy range varies with energy, being small at low energies and rising to a maximum just below the Fermi limit; the shaded area represents the total number of quantum states over the energy interval dE centred round E and, as this varies with E, the total area under the curve thus gives the total number of quantum states over the entire band. At temperatures other than 0 K the number of states per unit volume in the range dE is given by the well-established Fermi–Dirac distribution function:

$$N(E)\mathrm{d}E = F(E)\mathrm{d}E/[1 + \{\exp(E - E_F)/kT\}] \tag{3.7}$$

where $F(E)$ is the maximum available density of states corresponding to the case when all levels have been populated, and the factor $[\{\exp(E - E_F)/kT\} + 1]^{-1}$ represents the probability that a state of energy E is occupied at the temperature T K. The effect on the distribution of increasing the temperature above 0 K is shown on the diagram – the tail at the high energy end and a rounding off of the sharp 'head' should be noted; the states in the 'tail' beyond E_F belong, in the case of metals, to those which furnish electrons for thermionic emission, and the form of their distribution has undergone a marked change, resembling that for the Maxwell–Boltzmann, and displaying the same characteristic strong temperature dependence of the latter. The change in character of the energy distribution of the electron gas from F–D to M–B can be looked on as a direct result of the change in media in which the electrons find themselves; inside the metal quantum restrictions operate to determine both energy and density; once outside the influence of internal periodic fields and potential barriers of various kinds these restrictions no longer apply, and the electrons are free to take up any energy supplied by an external field.

When we consider the distribution applicable to impurity semiconductors, we must remember that the gap between the impurity level and the highest filled or lowest vacant levels may be quite small – of the order of the ionisation energy of the impurity atom embedded in a host lattice of high dielectric constant. The density of states in the conduction or valence band for a non-degenerate electron gas is then

$$N(E)\mathrm{d}E = F(E)\,\{\exp(E_F - E)/kT\}\mathrm{d}E \tag{3.8}$$

since the term unity in the denominator of equation 3.7 can be neglected, i.e.

$$N(E) \propto E^{\frac{1}{2}} \exp(-E/kT) \tag{3.9}$$

The electrons in the comparatively poorly populated levels of this class of solid thus obey a form of Maxwell–Boltzmann law, though their freedom is still limited to some extent by the restrictions imposed by quantum requirements of the lattice structure.

In the case of intrinsic semi-conductors and insulators, where the width of the forbidden gap may be quite large, an analysis shows that the Fermi level is approximately midway between the conduction and valence bands. The form of the function for the density of states will rise at the low energy end like that for a metal for each band; it will fall much more gradually on approaching the high energy end, and there should be no sharply defined limit as with metals; i.e. the edge should be rather broad. This is exactly what happens when the band structure is examined by the soft X-ray technique. In all cases the experimental values of the band width exceed those calculated on the 'free' electron theory; the more complicated the lattice the greater the difference introduced by using the 'free' electron approximation. The electron is therefore far from 'free' when dealing with these classes of material. The density of states in the bands can be simply stated for the energy interval E to $(E + \mathrm{d}E)$ as follows:

$$N(E)\mathrm{d}E \propto (E - E_{\mathrm{c}})^{\frac{1}{2}} \ldots \text{bottom of conduction band} \qquad (3.10)$$

$$N(E)\mathrm{d}E \propto (E_{\mathrm{V}} - E)^{\frac{1}{2}} \ldots \text{top of valence band} \qquad (3.11)$$

where E_{c} and E_{V} represent the potential energy levels of conduction and valence bands with respect to some zero level, say the vacuum level.

3.4 The Effect of Lattice Field: Brillouin Zones

The 'free' electron concept which has been used to explain the electrical and other properties of metals is only a first approximation. It was developed for the ideal case in which the valence electrons in the conduction band move through a region of the lattice in which the average field is everywhere practically zero; i.e. the electrons are not subject to acceleration except when the external field is applied. Such a view leads to a representation in terms of a box model whose profile in the one-dimensional case would consist of a base line – the conduction band – extending throughout the crystal and bounded at its ends by two lines at right angles to it, the potential boundaries, of height V, to the vacuum level, as illustrated in *Figure 3.3*.

The electron energy progressively increases from the bottom of the band upwards, and the proportion of kinetic to potential energy increases in the same direction; the topmost level corresponds to the Fermi energy E_{F} and represents the maximum energy in the band: the empty levels above E_{F} can, however, be filled by promoting electrons from the vicinity of E_{F} by various physical processes. The energy required to take an electron from the level E_F and place it just outside the material, at the vacuum level, is $e(V - E_F)$ where the bottom of the conduction band is taken as the reference level; this energy difference is known as the work function φ of the metal or solid.

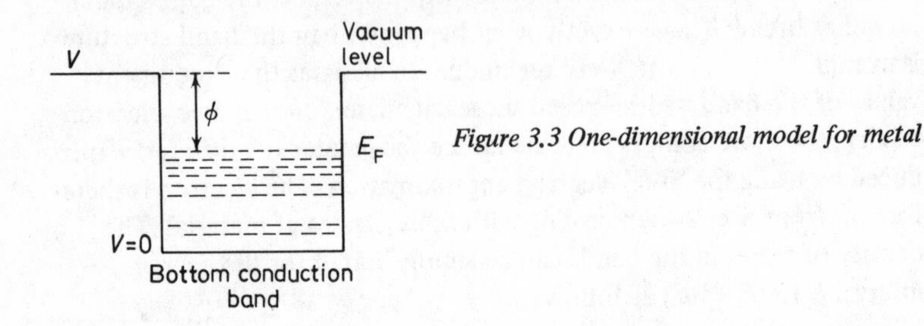

Figure 3.3 One-dimensional model for metal

The wavelength λ associated with the motion of the electrons between the potential boundaries is given, as stated earlier, by the De Broglie relation $\lambda = h/mv$ where h is Planck's constant and $mv(= p)$ is the momentum of the electron. Transposing gives

$$p = h/\lambda = hk/2\pi \tag{3.12}$$

where $k = 2\pi/\lambda$ and is known as the wave number and represents the number of crests over an angular distance of 2π. The kinetic energy of the electron is related to k by

$$E = p^2/2m = h^2k^2/8\pi^2 \tag{3.13}$$

so that a plot of E against k for this ideal case of 'free' electrons should give a smooth curve (*Figure 3.4*). In actual fact, the electron during its motion

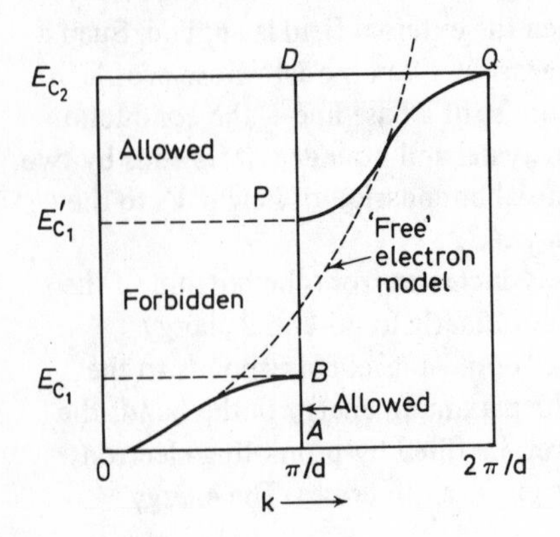

Figure 3.4 Brillouin zone formation

is bound to come under the influence of the periodic field of the lattice ions, and will be diffracted according to the Bragg law when $2d \sin \theta = n\lambda$ where d is the lattice constant and θ the angle of incidence. Taking a one-dimensional chain of atoms and perpendicular incidence, the electron wave will be reflected back in the direction from which it came when its wavelength satisfies the Bragg condition, i.e. when $\lambda = 2d/n$; there will thus be a critical value k_c of the wave number which satisfies $k_c = \pm n\pi/d$ (negative if the direction of incidence is reversed), and a corresponding critical energy E_{c1} for which no further motion of the electron in this direction through the lattice is possible. For $n = 1$, the value π/d defines the limit or boundary of the first conduction band or Brillouin zone; as the critical value $\pm k_c$ is approached, the energy states crowd closer and closer together, so that a small change in energy in this region corresponds to a very much larger change in k; the top of the band lies at the critical energy E_{c1} and this, as *Figure 3.4* shows, falls below the free electron value. The value k_c thus marks a transition between allowed and unallowed states; in the immediate vicinity of k_c the electron energy jumps discontinuously from the value E_{c1} corresponding to B, to the value of E'_{c1} corresponding to P. The energy interval BP represents a band of forbidden energy, hence, electrons with energies within the range B to P will be reflected back and will not be transmitted through this lattice.

Using a similar argument, a second Brillouin zone would begin at P where the energy is higher than that for a 'free' electron, and would end at Q, for which $k_c = 2\pi/d$; extension of these ideas allows us to conceive the complete energy system of the lattice as consisting of a whole series of alternate allowed and forbidden bands. There will, of course, be a similar curve to the one shown, a mirror image in fact for motion in the opposite direction, i.e. for negative k values, but this has been omitted from the diagram. When these concepts are extended to the three-dimensional case we find that the boundaries of the zones are determined by a whole range of values of k_c, one for every possible direction of motion of the valence electrons through the crystal, and the boundaries of the zones usually define complicated three-dimensional geometrical models. These can be constructed by making use of certain properties of the 'reciprocal lattice' of the original translation lattice; the surfaces of constant energy in k space can be mapped out and the energy gradient determined for every direction through the crystal.

The band structure thus reveals the presence and extent of the overlapping of the Brillouin zones. In the case of Na, with its single valence electron, the first zone is half filled, and there is some evidence of overlap of the second zone. With Mg, on the other hand, the first zone is completely filled, while the second, which overlaps it, has empty levels available which the conduction electrons can occupy: generally this overlapping of zones is a characteristic feature of the energy band system in all true metals, and the degree of overlap is related to the range of conductivity found experimentally.

The way the periodic field of the lattice affects the distribution of states is pictured schematically in *Figure 3.5.* The 'free' electron parabola

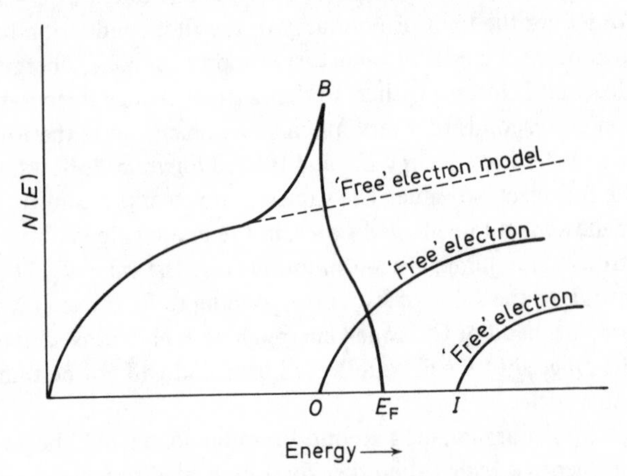

Figure 3.5 Effect of periodic field on distribution of states

is followed over the low energy portion of the curve, but there is a departure from it with increase in energy. The point B represents the boundary of the first Brillouin zone for certain directions in the crystal, and the curve shows both the crowding of states as the zone boundary is approached, and the decrease in the density of states for other directions as the energy corresponding to B is exceeded. The point E_F represents the maximum energy in the band when all possible directions in the crystal have been explored, and the second zone is shown as beginning at O when the overlap occurs.

The band system of insulators will, of course, consist of filled zones with forbidden regions or gaps, indicated on the figure by the interval before the second zone, marked I, is reached. These are usually so wide that it is not possible under ordinary conditions to promote electrons to the empty levels in the bands which lie above the gaps. In semi-conducto the energy difference $(I - E_F)$ is small enough to be bridged by electrons, if they are thermally excited or can absorb the necessary quantum energy from a source of external radiation. The structure in the zone which results from the effect of the periodic field of the lattice is an important current topic of research on bands in metals and alloys. The experimental confirmation of the existence of the Brillouin zones will be brought out later when interpreting the results of soft X-ray studies.

Chapter Four

THE TECHNIQUE OF SOFT X-RAYS

4.1 Introduction

As the majority of soft X-ray spectra fall in the range from about 50 Å to 500 Å, i.e. between the hard X-ray region and the extreme ultra-violet, it is impossible to find natural crystals which can diffract them. The Bragg condition $2d \sin \theta = n\lambda$, which relates the wavelength to the lattice constant d, requires lattice spacings (for perpendicular incidence) of 25 Å for a wavelength of 50 Å; the maximum possible with bent mica and gypsum crystals, of radii of curvature 25 cm, is about 19 Å. To get over this difficulty recourse is made to the concave diffraction grating, usually ruled with 30 000 lines per inch for this range of spectroscopy, i.e. with a grating constant of some 1200 Å. The soft X-rays are made to impinge on this at nearly grazing incidence for two reasons: (a) the reflection coefficient is high enough even at the lower wavelengths, and improves as λ increases; (b) increased dispersion results from reduction in angle of incidence.

The condition for reinforcement on diffraction is readily derived from *Figure 4.1* which shows the incident and diffracted rays making angles of θ and $(\theta + \alpha)$ with the surface respectively. Since the difference in path lengths must be an integral number of wavelengths, we have

$$PS - RQ = n\lambda = d\{\cos \theta - \cos(\theta + \alpha)\}$$

$$\doteqdot d\{1 - (\theta^2/2) - 1 + (\theta^2 + \alpha^2 + 2\alpha\theta)/2\}$$

for small θ

$$n\lambda = (d\alpha/2)(\alpha + 2\theta) \qquad (4.1)$$

As the incident angle θ is fixed by the initial geometry, and α can be measured from the position of the diffracted beam, the wavelength range of the band, and hence the energy, can be determined. The dispersion of the instrument, defined in terms of $d\alpha/d\lambda = n/d(\alpha + \theta)$, shows an increase as θ is reduced; it also increases with n, the order of

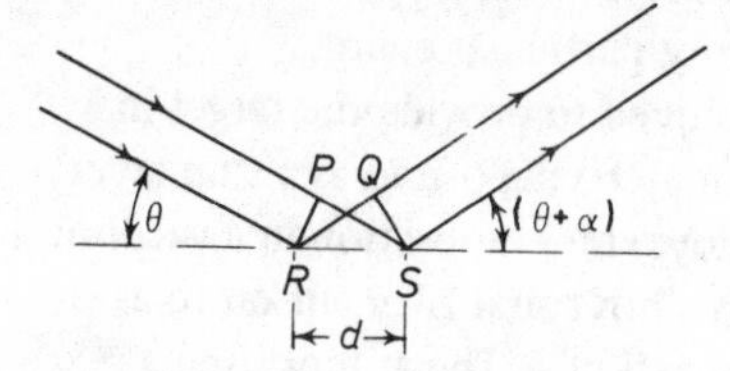

Figure 4.1 Diffraction at a plane grating

the spectrum, so that it is sometimes advantageous to work with high order spectra on this account, particularly when investigating details of structure over the range of wavelengths in the band.

4.2 Sources and Emission Techniques

The soft X-rays are generated by electron bombardment of an anode of the material which is being investigated; in some cases, for example C, it is necessary to paint (or strap) the material on to a copper or other metallic target to a sufficient depth in order to simulate bulk properties: both the electron current and the energy of the beam are suitably controlled. Under the bombardment the electrons in the target material are knocked out of one or more of the K, L, M, . . ., shells by the incident electrons. Since little or no overlap of wave functions occurs in the inner levels of the atoms when at their equilibrium distance in the solid, the levels are comparatively sharp in relation to those making up the valence band; the subsequent transitions of electrons from each and every level in the valence band into the vacancies created in the sharp inner levels by the bombardment are governed by the selection rules applicable to atomic transitions. Their number at any energy depends on the density of electrons in quantum states at that energy which can make the transition to the vacant inner states. Each electron filling a vacancy emits a quantum of radiation, of energy $h\nu$, corresponding to the difference in energy between its initial and final state; the emitted radiation thus reflects the range of energy in the band. The density of quanta liberated in such processes corresponds approximately with the density of electrons making the transition (modified by the transition probability), so that the intensity recorded on a photographic plate or counter is thus a measure of the density of states. If the response of the detector is known or can be explored over the whole wavelength range, it is possible by such a technique to get a reasonably faithful picture of the way the quantum states are distributed in the energy band, and hence the density of states, i.e. the number of states per unit volume per unit energy range centred round any particular energy.

Various techniques have been employed to provide the target in a suitable form; for example, the solid may be deposited as a thin layer (for pure metals and also for some alloys) by evaporation in a vacuum from a quartz or carbon boat or from a hot spiral filament on to a suitable rigid, water-cooled metal anti-cathode. The author used a hollow copper block of square section 1 cm^2, as shown schematically in *Figure 4.2(a)*. Such a deposited layer, about 1000 Å thick, can be

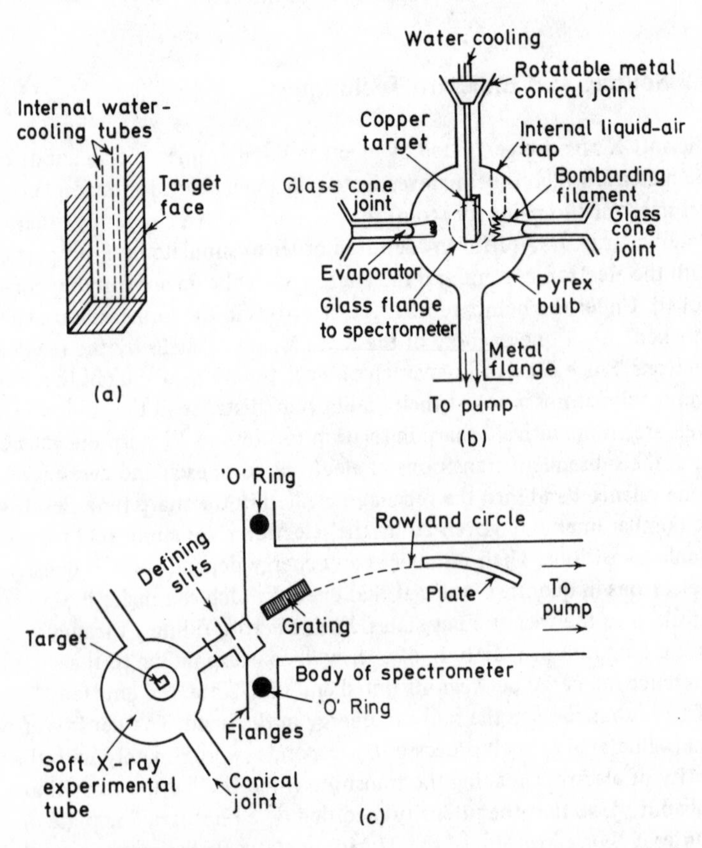

Figure 4.2 (a) Water-cooled copper target, (b) soft X-ray experimental tube, (c) the complete spectrometer

renewed from time to time by evaporation, so that contamination of the surface can be kept to a minimum; a film thickness of this order is sufficient to simulate all the X-ray properties of the material in bulk. In the author's arrangement the target, evaporator and electron gun were mounted inside a 1 l Pyrex flask (which contained an internal liquid-air trap) as shown in *Figure 4.2(b).* It is necessary in the case of some alloys to cut a flat sheet to size and tie it with wire to the face of the target; occasionally it is possible to rivet it to a sheet of nickel which is then 'soldered' to the target face.

4.3 The Spectrometer Layout

The X-rays emitted by the target are taken off in vacuum to avoid absorption, at a small angle, (~ 6°) to its face, as shown in *Figure 4.2 (c),* through defining slits in the spectrometer until they are incident at almost grazing incidence on the concave grating whose radius of curvature defines the diameter of the Rowland circle illustrated in *Figure 4.3.* To give some idea of dimensions, the author used a source

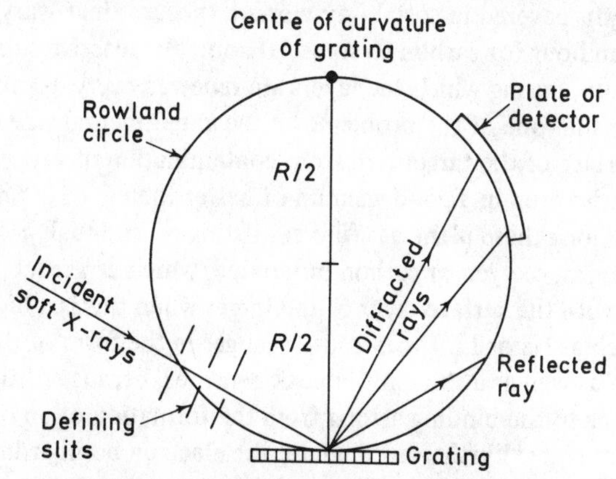

Figure 4.3 Focusing using the Rowland circle

slit of 0.03 mm, through which the radiation from the target was led into the spectrometer, followed by an intermediate slit, 1.5 mm wide, and a final slit of variable but controlled width, ~ 0.7 mm nearest the grating. With this arrangement it was possible to ensure that about 1.5 cm of the grating was irradiated by the X-ray beam, since this width of the grating was the optimum necessary for the geometry used. It is usual to examine the focusing and dispersing quality of various regions of the grating, if this is not uniformly good. The necessary condition for focus of the diffracted beam requires that all three – spectrometer slit, point of contact with the grating (which is, of course, tangential to the circle at that point) and the position of the detector – should lie accurately on the Rowland circle. When this condition is correctly fulfilled the diffracted rays are automatically in focus on the other side of the normal, over an arc of a circle. The greatest possible care must be taken when setting up the system to ensure correct positioning; here, it is helpful in the first instance to observe the optical reflection by the grating of a pilot lamp placed behind the first slit, and to correct for tilt, sharpness, etc., by trial and error methods before examining provisional exposures. A final test should show all the different orders of the spectra in focus over the plate or detector, and quite sharp against any background radiation which may be present.

The whole spectrometer, together with the X-ray source, should be kept continuously evacuated, day and night, to a pressure not exceeding 10^{-5} mm Hg. The special photographic *Q* plates made by Ilford, with thin emulsion and high silver content, respond satisfactorily to the range of wavelengths covered in soft X-ray work. Exposure times vary from about half an hour for carbon to some 70 hours for reactive substances like potassium, during which the layers are renewed every 10 min. It will be clear that one of the problems on the experimental side is to keep the surface of the target free from contamination of various kinds which can arise even in a good vacuum of better than 5×10^{-6} mm Hg. Foremost among these is the gas film consisting of 'residual' gases, mainly nitrogen, oxygen or carbon monoxide, which can react chemically with the surface layer of the target when this is a highly active material such as Na or Li. It can cause changes in the form of the distribution as well as a shift in the peak densities, because of the modifications to the binding arising from the formation of an oxide or other surface layer. Furthermore, under the electron bombardment of the layer in the presence of an electric field, a certain amount of electrolytic dissociation in a direction perpendicular to the target surface takes place, over and beyond the depth of penetration of the electrons into the material. In addition, dissociation products of the low vapour-pressure vacuum pump oil may cause the deposition of fine carbon particles in patchy regions over the surface of the target; the possibility of their diffusion into the lattice of target atoms also exists. A series of carbon bands, strong or weak depending on the concentration, may thus appear in many orders superposed on the band of the material of the target being investigated. This can lead to complications in interpretation, although the presence of the carbon bands can be helpful in providing a wavelength calibration for the plate.

The obvious remedy for this state of affairs, assuming the well-established principles of vacuum technology have been applied, is to use the minimum possible energy to the target, coupled with an exceedingly fast detecting technique such as the incorporation of a Be–Cu photomultiplier into the vacuum spectrometer. The latter can be made to scan the diffracted beam and so pass on the information about intensity, i.e. number of quanta, to an amplifier and counter; thus we can have available a visible trace of the distribution on a recorder actually outside the vacuum apparatus. This improvement enables exposure times to be cut down to a few minutes' duration; the new technique of recording has so far given the most accurate and reproducible results. A further and not insignificant advantage accrues from the direct measurement of intensities, since the photographic exposure containing the information on band form, structure and relative intensity has to be analysed with the help of a microphotometer, which has to be previously calibrated, in order to record the density variations of the plate; these need further corrections, for example, background, before reliable conclusions can be drawn from the traces. It is worth mentioning that care in interpretation is necessary with all systems of recording in the region of wavelengths which are absorbed by the silica of the grating itself.

4.4 Photoelectric Recording

Future technique in the soft X-ray field will undoubtedly concentrate on the development of photoelectric recording, and it would perhaps be of interest to describe briefly the design of this type of spectrometer as introduced by Skinner at Liverpool (see *Figure 4.4*). Here the detector is mounted at one end of a radial arm, the other end of which is accurately pivoted at the centre of the Rowland circle, radius 50 cm, so that the diffracted radiation can be picked up over an arc extending through one-quarter of the circumference, if necessary. The signal from the final dynode of the 15-stage Be–Cu photomultiplier is taken out of the vacuum system to a leak-free terminal block (not visible on *Figure 4.4*) and to an electrostatically screened pre-amplifier, then through the main d.c. amplifier (gain 10^4) which amplifies the average current output and feeds it to a pen recorder which traces out the distribution on a chart. This is actually proportional at each point of the trace to the density of quanta in the band at that point. As an alternative method of detection, it is possible to amplify and count the pulses coming from the target at selected locations of the scanning arm, as it is made to cover the band. The usual statistics are applied to make the necessary corrections required by the scaling unit; the accuracy can of course be increased by making the time of scanning longer, but there is a limitation here because of contamination and a compromise is usually sought.

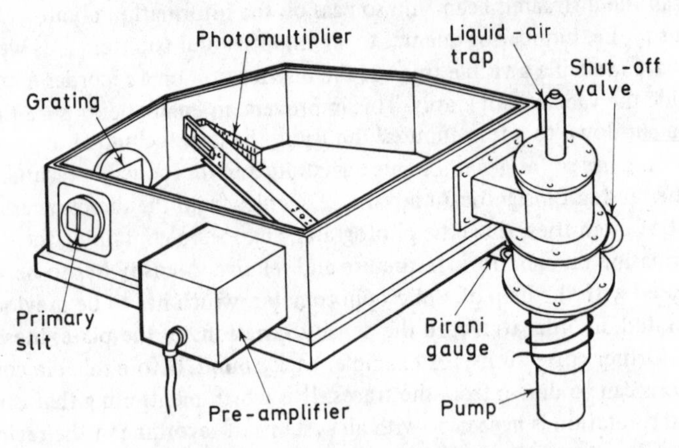

Figure 4.4 Photoelectric recording

This new technique of rapid recording is particularly valuable when used with materials which may change their structure significantly because of surface effects or through the bombardment; it is estimated that artificially introduced lattice defects can be cut down by a factor of 10 in this way. To achieve success one must use the best vacuum techniques, sound mechanical design and construction, and a 'takeover' of well-established counting techniques developed for nuclear research. The method also opens up the possibility of developing a new 'tool' for the study of gas or other contaminations on the surface of metals, semi-conductors and insulators.

4.5 Absorption Techniques

The above remarks apply with equal force to investigations involving the absorption coefficients of solid materials. As is natural to expect, most of the published results have been obtained using the photographic method. In absorption studies the experimental procedure involves the preparation of a thin film absorber of the material, of thickness 500–1000 Å, by evaporating from a filament, preferably *in vacuo*, on to a small metal grid or small hole in a plate. The aim is to produce a continuous and homogeneous absorbing foil; in some cases it is necessary to deposit the evaporated film on a 'zapon' structureless substrate which is placed across the aperture. Experience has shown that such a substrate can introduce unexpected features into the form of the absorption curve beyond the absorption edge; great care is necessary to avoid complications arising from interfacial contamination, for example, diffusion at the boundary between substrate and evaporated layer. Ideally the absorbing film should be placed on the source side of the first slit; if this is done, difficulties arise which affect the continuity of the film, for example, stray electron bombardment, back reflection of the electrons from the slit and glasswork, the charging up of the film, contamination and attack of the film surface by vapours which exist in the X-ray tube itself. In practice, it is found essential to place the absorber behind the slit; such a position shields it electrostatically and avoids most of the complicating factors which lead to surface contamination and physical rupture of the film. Great patience and first-class technique are needed to procure sound and reproducible results in absorption work.

In order to eliminate spurious interference from the interfacial layer, a comparison is made between two absorbing films of different thicknesses deposited on the same substrate at the same time; the thickness being calculated from the mass of material evaporating per second at the temperature of the 'oven' or being measured directly, using a Michelson or Fizeau interferometer. Double exposures are made for each film; those for the thin film differ in time by a factor of two, whereas those for the thick films are increased by a factor F, so that approximately the same number of X-ray photons reach the photographic plate for all exposures. If the plausible assumption is made that over the range of wavelengths covered by the band, the absorption coefficient is independent of the wavelength, it is possible to work out the response curve for the plate. On repeating this procedure over different ranges of wavelength we are able to construct a curve relating the absorption with the wavelength; the coefficient μ for the film of thickness equivalent to the difference between the thick and thin absorbers is found from the usual relation $I = I_0 \exp(-\mu t)$.

As regards standard wavelength sources, it is of course essential that the radiation should be intense enough to penetrate the absorbing film. For this reason it is usual to employ the following:

1. A vacuum arc between copper or silver electrodes, which furnish atoms in an ionised state and thus provide a suitable range of spectral lines; the rate of 'breakdown' is controlled by an air gap in series with the main gap.
2. A discharge tube of special design in which a rare gas like He or Ar at a pressure of about 1 mm Hg flows through a quartz capillary and is excited by the passage of a condensed spark through it.
3. Background radiation from the electron bombardment of a suitable metal target, for example, Au, gives radiation over the range required for *K* absorption with B_2O_3 films; its intensity can usually be varied to suit the particular type of absorber.

Synchroton radiation has been used to provide higher resolution and stronger continuum sources, but is obviously out of the reach of most research institutions.

Chapter Five

INTERPRETATION OF SOFT X-RAY SPECTRA

5.1 Introduction

In this chapter evidence is presented which justifies our confidence in the theoretical deductions derived from the band theory of solids sketched in broad outline earlier. This has supported the classification of solids based on the three main types, metals, semi-conductors and insulators. The techniques of soft X-ray emission spectroscopy when applied to the metals supply a great deal of important knowledge about the filled energy levels which terminate on the high energy side at the Fermi limit; straightforward considerations lead us to expect that this emission edge should coincide with the absorption edge, and this is found to be so. On the other hand, in the case of both semi-conductors and insulators theory suggests and experiment confirms a shift in the absorption edge in the direction of shorter wavelengths, i.e. higher energy photons than those corresponding with the emission edge are required to account for this difference; the amount by which the edge is displaced must correspond with the width of the energy gap between the full valence band and the empty conduction band.

The information derived from a study of 'emission' spectra complements that obtained from 'absorption' spectra; correct interpretation of the results enables us, in theory at least, to build up a composite picture of the whole band system of the solid. In practice, while the main features of the band system can be reasonably well established, there are difficulties when dealing with the details. An approach from the theoretical side has been made to elucidate the structure of metals, alloys and chemical compounds, and some of the results can be put to the test using the soft X-ray method. This whole field is therefore of much interest to metallurgists and chemists as well as to physicists, both experimental and theoretical. Matters of interest and pertinent questions concern the following:

1. The width of the band of the component metals in the alloyed state compared with that of the pure metals.
2. The effect of alloying on the high and low energy parts of the band.
3. The corresponding degree of overlap of the Brillouin zones.
4. The change in electron concentration round component atomic centres.
5. The development of special structural features and their dependence on composition.
6. The bands of magnetic alloys above and below the Curie point.
7. The shifts in the positions of the peak (maximum density of states) in chemical compounds, compared with their positions in the pure elements; this gives information on the heat of formation of the compound as well as on the binding.
8. The band width in oxides, reflecting structural changes due to chemical bonding.

5.2 The Emission Process

Before going on to present a series of representative emission spectra obtained by the experimental method, it is necessary to look a little more closely into the background of changes brought about by electron transitions between the valence band and the inner levels. The simplified analysis presented earlier leads to a value for the density of states at any energy value; it was based on quite general assumptions about energy and momentum states in a solid without reference to any particular lattice structure, and is therefore likely to need modification on this account. Furthermore, the superposition of bands and the partial occupation of the level system by states with different symmetry properties complicates the interpretation of results. However, the selection rules applicable to X-ray spectra are obeyed for the radiative transitions from the valence band to the sharp inner levels leading to the emission of the soft X-ray photon. This means that the intensity recorded by the detector of the spectrometer only reflects those states in the band which, according to these rules, can make the transition.

The presence of other states occupying levels in the same band are consequently not detected under these conditions; they may, however, be sorted out when they participate in the other types of transition; for example, one might expect a form of the intensity distribution over the band to remain essentially the same whether the final, single state was K or L. This is not found to be so – the transitions s → p and p → d are allowed, so that where the initial state is p, as in spectra involving final end points on L_2 and L_3 (designated L_2L_3 spectra), the transitions only involve s and d states of the valence band; where it is s, as in K spectra, the intensity distribution reflects states of p symmetry, since only p → s transitions are allowed.

In order to facilitate our understanding of the form of the radiation intensity over the range of frequencies in the emission band, it is necessary to consider the factors involved in the process of emission. These come under three headings:

1. The energy of the initial and final states.
2. The wave functions of the electrons in these states.
3. The numbers of electrons involved.

The probability of an emission event taking place varies with the positions of the levels concerned and the level structure. The intensity of the emitted radiation is proportional to the number of electrons which make the change, the square of the energy emitted (and hence ν^2) and a complicated function F involving the transition probability and the gradient of the constant energy surfaces in k space. We may therefore write that the intensity of the radiation $I(E)$, measured in terms of the energy emitted per second per unit solid angle falling on the detector, varies as $\nu^2 N(E)$, i.e. as $\nu^2 E^{\frac{1}{2}}$ (where ν is the frequency of the emitted radiation) modified by the factor F which depends on the final state. When dealing with K spectra and for energy values near those at the bottom of the band, $F \propto E$, whereas for L type spectra F is independent of E. This leads to the following relations:

For K bands

$$I(E) \propto \nu^2 E^{\frac{3}{2}} \tag{5.1}$$

For L bands

$$I(E) \propto \nu^2 E^{\frac{1}{2}} \tag{5.2}$$

It follows therefore that where overlap occurs, the recorded intensity variation over an emission band is not an exact representation of the actual density of states over each energy range; the actual density of levels corresponding to any energy E is therefore made up of the sum of contributions arising from electrons of s, p, d . . . type wave symmetry, i.e.

$$N(E) = N_s(E) + N_p(E) + N_d(E) \tag{5.3}$$

If we apply the selection rules, we may say that K spectra involve only p electrons; while L spectra represent a combination of s and d electron states, which almost certainly for low energies in the band consist predominantly of s electron transitions, since the contribution from d states under these conditions is very small; at higher energies the intensity distribution reflects contributions from both states.

5.3 Emission Spectra

5.3.1 Monovalent metals

The importance of the experimental method lies in our ability to get an overall picture of the form of the band, with reasonably accurate values for band width and edge location, together with special features which develop under certain circumstances; these may help in supporting and extending theoretical developments.

The next stage is to look at the recorded band patterns for some metals, semi-conductors, insulators, simple alloys and chemical compounds. The object is to attempt an analysis and present general characteristics of the main types of systems of energy levels.

We first take the body-centred, cubic alkali metal, Na, a good conductor, with one 3s valence electron per atom, so that the first Brillouin zone is half full; the band structure actually involves overlap of the 3s and 3p levels. As regards the valence electrons, we should expect that since the k vector for these is far from the zone boundary they would behave as 'free' electrons in the Sommerfeld sense; this implies that the highest energy level should be sharply defined, with the consequence that the 'head' of the band in this region should exhibit a steep fall in the density of states, cutting the energy axis at the Fermi value. *Figure 5.1* shows this clearly for the L_{23} band, for which the breadth of the edge is exceedingly small – some hundredths of an electron volt. No step or doubling of the edge is to be seen, which, if present, would indicate the existence of L_2 states, although the L_2 and L_3 levels which only differ in spin, overlap, their separation of 0.21 eV is too small to be resolved with the instrument used.

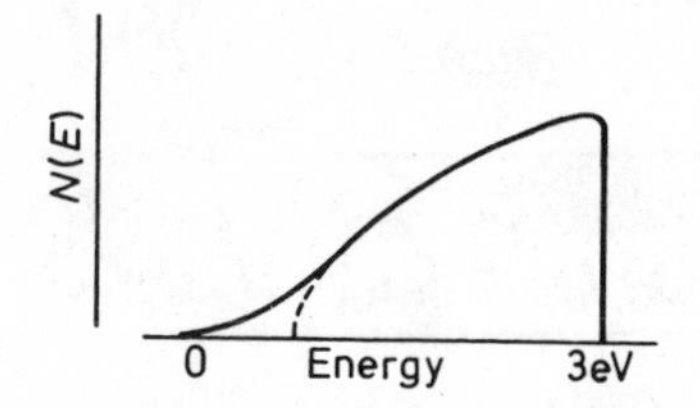

Figure 5.1 Experimental distribution of states: Na L_{23} *band*

The low energy end of the band shows a curve of density of states $N_{(s+d)}$ which should vary as $E^{\frac{1}{2}}$, but experimentally conforms more nearly to an $E^{\frac{2}{3}}$ law. The reason for this is the superimposed 'tailing', arising from the existence of non-radiative transitions which take place within the band when electrons go from higher to lower levels; this results in the release of 'Auger' electrons from the top of the band. When this is allowed for, the empirical band width obtained by extrapolation comes out as 3.0 ± 0.2, and this agrees with the value 3.2 eV calculated from the Sommerfeld formula $(h^2/8m)(3N/\pi)^{\frac{3}{2}}$.

It can be seen that the parabolic form predicted from the Sommerfeld distribution is, on the whole, well maintained for Na; the use of the 'free' electron model therefore inspires confidence in its application in spite of its neglect of the crystal structure. This last point is emphasised by the results on Li, a member of the same family. Here the fall at the high energy end is much more gradual than in the case of Na, and the width of the edge may approach 1 eV; this has been attributed to changes in crystal structure with the temperature of the Li target under the electron bombardment, quite apart from instrument limitations.

5.3.2 Divalent metals

Here, with two valence electrons, the band form shows several new features. As an example, the distribution for Mg is shown in *Figure 5.2*; the two valence electrons obviously fill the first Brillouin zone, and the second zone must overlap the first in order to confer conducting properties on the material. The peak marked P must consequently represent the maximum density of states, and corresponds to energy contours in k space which just touch the boundary of the first Brillouin zone for certain directions in the crystal; the $N(E)$ curve then falls along PDF as higher energy contours are included for other directions, until we meet the maximum energy for the zone at F.

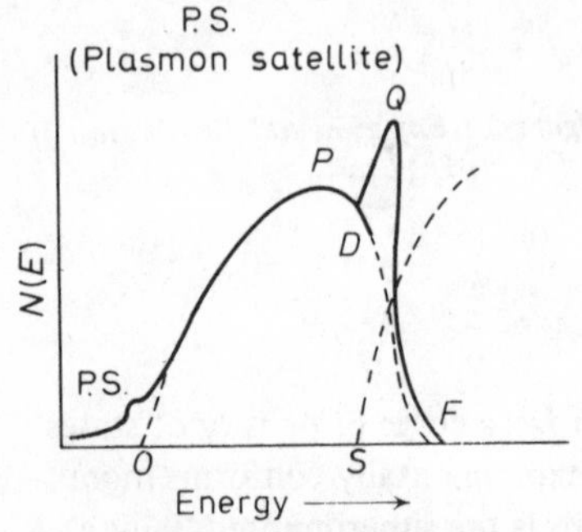

Figure 5.2 The Mg L_{23} *band distribution*

The overlap with the second zone begins at S, at an energy which is much less than the highest in the first zone; the superposition of the two densities causes the depression at D and the subsequent rise to the spur at Q, while the abrupt fall to the energy axis defines the Fermi limit at 7.3 eV, and also reveals the existence of the L_2 and L_3 edges (resolved in the original photograph). On the low energy side the curve should rise as $E^{\frac{1}{2}}$ but, as with Na, there is 'tailing', and the experimental distributions make a better fit with an $E^{\frac{2}{2}}$ law; when this is taken into account and the empirical rule applied, the band width of 7.3 eV agrees with the calculated value. It is found that for the elements of the first two groups of the Periodic Table the calculated values are generally somewhat smaller than those observed for the band width. Although this is the first case we have discussed of the visual existence of the Brillouin zone boundaries in the distribution, all interpretations made on the metal bands fit in with such a description; i.e. we have here excellent confirmation and direct evidence of the presence of these zones in a metal: this places the theory on a really sound basis.

It is interesting to record that both K and L spectra yield the same value for the band width of Mg, in spite of the fact that both the experimental and theoretical forms of the distribution are different for s and p excited states because of differences in transition probability. The true band width may lie on either side of the calculated value, since the $N(E)$ curve up to the Fermi limit must define the total number of electrons in the band; the shape of the distribution and the exact position of the Fermi edge are therefore sensitively affected by this condition.

5.3.3 Trivalent metals

We expect somewhat more complicated forms for the distribution of states in trivalent metals, and indeed this is so as can be seen from *Figure 5.3(a)* which represents the band of Al (with the L_2 contribution completely removed). Here the first Brillouin zone is filled by the s states of the M shell, while the second is only partially full and overlaps the first. This means that, as the curve shows, the constant energy contours in k space first touch the boundary of the first zone at P, while the dip at D defines the lowest energy in the overlapping second zone; this, like the low energy stage of the first zone, rises theoretically as $E^{\frac{1}{2}}$. Superposition of contributions from both zones gives the complicated resultant $N(E)$ curve, and the sharp termination at the band 'head' defines the Fermi level; this emission edge corresponds exactly with the absorption edge, though its width varies with temperature.

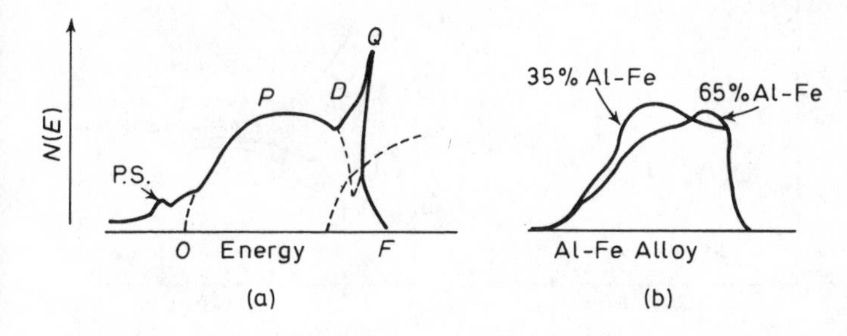

Figure 5.3 (a) The Al L_3 *band distribution, (b)* Al *band in* Al–Fe *alloy*

The original trace of the L_2L_3 band shows both L_2 and L_3 edges quite well resolved, though there is no trace of the L_1 edge on the microphotometer record. The empirical band width, assuming a linear extrapolation at the low energy end, comes out at 13.2 ± 0.4 eV, as compared with the calculated value of 11.9 eV; the agreement to about 10% can be considered satisfactory, since the latter neglects to take into account both the relative importance of s and d functions, particularly at the higher energies and their involvement in the transition probabilities. The increase in band width as the number of valence electrons per atom increases is in accord with simple theory, the larger the number of electrons per unit volume the higher the energy of the highest state.

5.3.4 Semi-conductors

As an example of a semi-conductor, *Figure 5.4(a)* shows the form of the valence band distribution for tetravalent Si. Note the striking difference as compared with metals below the high energy end, and the absence of Brillouin zone overlap; this last is attributed to the complete filling of the zones with two electrons per level. The absence of an abrupt energy fall at the high energy end characterises this class of materials; in the case of Si the edge extends over a range of at least 1 V (it is just possible to see the L_2L_3 edge resolved in the original microphotometer trace). Furthermore, there is no variation of band shape or edge width as the temperature is raised or lowered above or below room temperature. This kind of behaviour is also exhibited by semi-conductors like B and C, and one could form a classification of them by using this specific property.

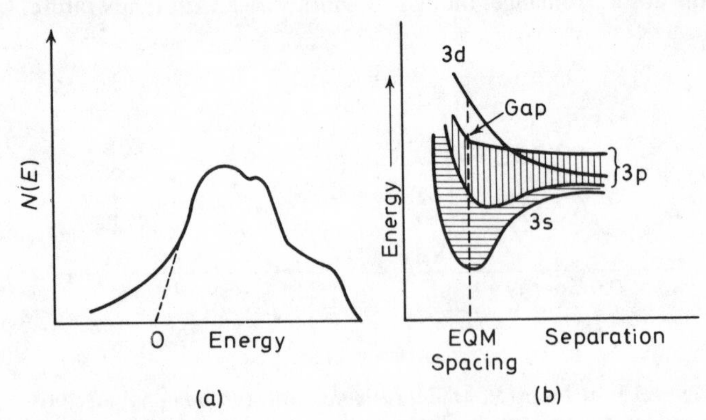

Figure 5.4 Semi-conductor: (a) Si: L_3 *band, (b)* Si: *energy band system*

The band scheme for Si shown in *Figure 5.4(b)* has a gap about 1.2 eV wide at the equilibrium distance in the solid. The existence of the 'gap' can be inferred from an absorption experiment; the absorption edge appears at a shorter wavelength than that of the emission edge, and the difference expressed as an energy represents the width of the forbidden gap. A simple experiment can be set up to show that silicon absorbs strongly in the infra-red at a wavelength of 1.03μ, so that the photon energy required to take an electron across the gap is 1.2 eV; confirmation of this value comes from a measurement of the slope of the conductivity/temperature curve for the specimen; this gives the gap width and is in good agreement with this value. As regards the band widt the empirical values for non-metals are invariably larger than those calculated, for example, for Si 18.2 ± 0.5 eV and 12.7 eV respectively; this difference arises from the presence of 'false' band ends on the high energy side, and these are bound up with certain complications introduced by the lattice structure and will not be pursued further here.

5.3.5 Insulators

Ideas on the band structures of insulators come from an extension of those discussed for semi-conductors, for a distinguishing difference between these classes of solid is merely one of 'degree', affecting the width of the forbidden gap. Hence we expect full bands, no overlapping of zones and broad edge widths; indeed, these characteristics are typical with the insulator diamond and its allotropical modifications in the form of graphite. Both forms yield the same empirical band width (33.3 eV) though there are differences in detail arising from structural variations; it is evident from *Figure 5.5* that they follow each other as

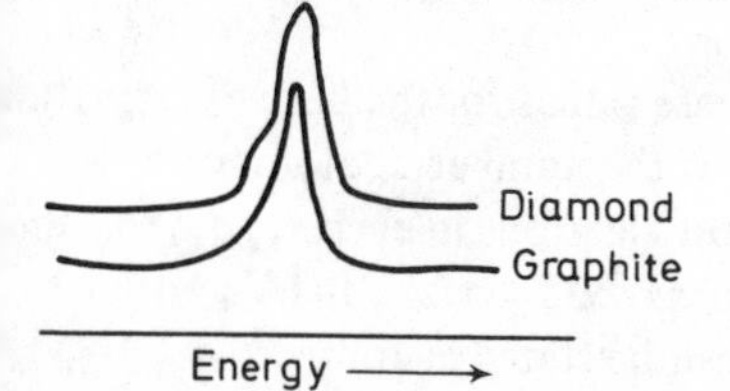

Figure 5.5 Comparison between K *bands of diamond and graphite*

regards general shape, and have false band ends which extend over some 15 V. The layered structure of graphite, consisting of planes containing C atoms arranged hexagonally, obviously makes for a considerable degree of asymmetry; this is reflected in the comparatively good electrical conductivity along the planes, and a smaller electron concentration per unit volume of the material and, hence, a smaller calculated value for the band width (21.9 eV). It is interesting to note in passing that it is possible to excite the carbon K_α spectrum in eight orders by electron bombardment of the cooled target on which an oil film (from the pumps) has condensed; this ubiquitous spectrum can be used as a secondary standard of wavelength for calibration purposes.

5.3.6 Chemical compounds and alloys

The formation of chemical compounds and alloys is the result of an exchange of valence electrons between the constituents; the resulting compound may take on a new character depending on the degree of exchange, i.e. on the proportions of the constituents. In this way it is possible to change the bonding from ionic, say, to covalent, for example, excess and deficit semi-conductors; changes of this kind involving new configurations of electrons and densities are reflected in both the form and distribution of the density of states as revealed by the soft X-ray spectrum of the compound. The components, now in the form of ions or perhaps in solid solution, show marked changes in distribution as compared with their pure form; as an example, the Mg component in MgO has a K band which has lost its edge, is symmetrical about the peak and is displaced towards lower energies compared with its position in pure Mg metal. This means that the potential energy of the electrons responsible for the binding has increased; the tighter binding must be associated with the release of energy, i.e. the evolution of heat, on formation of the compound. As regards the other component, the oxygen peak is found displaced towards higher energies, and thus corresponds to a decrease in potential energy, a looser binding and a heat of absorption on formation.

It is possible to calculate approximate values for the heat of formatio of the compound, given a knowledge of the number of atoms participating n_x, n_y; the number of outermost electrons in each e_x, e_y; togethe with the direction of the shift in the peak $\pm\Delta E_x$, $\pm\Delta E_y$ in eV. An empirical equation connecting these can be stated as follows:

$$n_x e_x \Delta E_x \pm n_y e_y \Delta E_y = Q/23.05 \tag{5.4}$$

where Q is the heat of formation in k cals/g mol. *Table 5.1* shows the extent of the agreement with the thermochemical data.

Table 5.1 COMPARISON OF APPROXIMATE HEAT OF FORMATION OF COMPOUND USING EMPIRICAL BAND PEAK SHIFTS, WITH THERMOCHEMICAL DATA (DAS GUPTA)

Compound	ΔE_x (eV)	Q_x (eV)	ΔE_y (eV)	Q_y (eV)	Q (eV)	*Q thermochemical* (eV)
MgO	4.5	+9	0.50	−3	+6	6.5
Al_2O_3	4.5	+27	0.50	−9	18	17
SiO_2	3.9	+15.6	0.50	−6	9.6	9
$AlFe_3$	2.0	+6	0.24	−4.8	1.2	0.5
SiC	0	0	0	0	0	0.05

Note that the numbers of outermost electrons used in the calculations are: Mg, 2; Al, 3; Si, 4; O, 6; Fe, 8. Alternatively it is possible from a knowledge of the thermochemical data to predict the direction and amount of shift, and to account for the preference of the transferred electrons for their new environment.

As regards simple alloys, the least complicated from the theoretical point of view is a dilute solution of one metal in another, for example, a few per cent of Al in Cu. Here the aluminium component loses its unique shape and its sharp emission edge completely disappears. This is explained by considering such small concentrations of Al to act as impurity centres in the Cu lattice; the distribution shown by the soft X-ray method is therefore characteristic of the localised energy states associated with the environment in the vicinity of the impurity atom. This, of course, is very different from that of pure Al. Under these circumstances the electron contribution from impurity atoms is negligible compared with that from the host; one may conclude that in the case of such weak alloys the impurity atoms take little or no part in the electrical conductivity of the alloy as a whole.

With concentrated binary alloys each component tends to preserve its own band width in spite of wide variations in composition; hence the band width is not structure sensitive in alloyed states, and our expectation of a possible common valence band cannot be realised. With the exception of the transition metals as well as Cu and Zn, the shape of the bands defining the density of states can change markedly as the concentration of the components is made to vary, though edge width and displacement changes are generally small; an example is shown in the author's curves in *Figure 5.3(b)* for the changes of form of the Al band as the percentage of Al in an Al–Fe alloy is varied. The change in shape on alloying is clearly the result of electron exchange as between the two types of atoms; the effect of reduced concentration appears to enhance the existence of low energy states.

The field of alloy structure as opened up by the soft X-ray techniques is a most important and growing one; both the chemist and metallurgist are directly involved in it, and the theoretical physicist can find a large number of problems, the solution of which would help considerably in the intricacies of interpretation. An exact and successful explanation of the distribution shown by commercial alloys would enable us to understand why and to what degree changes in electrical, thermal, mechanical, chemical and magnetic properties could be made to occur by suitable combinations of the components.

5.4 Special Features of the Bands

It is of interest ot point out some of the unexpected features found from examination of the experimental distributions, in particular:

1. The existence of 'tailing' at the low energy end of the band.
2. The recently observed 'plasmon' satellites in the tails of the K band of Be and L_{23} bands of Mg, Al, Na.
3. The departure from the theory of the intensity ratios as between the different ionised levels.

These points will now be discussed briefly in turn.

The theoretical rise of $N(E)$ with E at the low energy end of the band should follow either the parabolic or three-halves law; in both cases this requires that the curve should pass through the origin or the zero energy assigned to the bottom of the conduction or valence band. The experimental results particularly for L_{23} spectra of metals show considerable tailing at this end of the band, and is quite noticeable in *Figures 5.1* and *5.2*; although this 'tail' was responsible for only some 4% of the total intensity in the band, nevertheless it was so marked a feature of such spectra that it puzzled early workers. A satisfactory explanation was put forward by Skinner who pointed out that 's' and 'p' states in the band can be considered as completely merged; this means that no transition involving radiation is possible between levels within the band. However, a vacancy arising at the bottom of the band through a radiative transition to an allowed lower state could be filled by an electron originating at the top of the band, and this gives rise to a radiationless transition in which the surplus energy is imparted to an electron at the top of the same band. In essence this class of interaction can be labelled Auger transition, and it enables the liberated electron to enter one of the empty levels above the top of the valence band.

A sequence of this kind results in the shortening of the life time of the 'hole' created by the removal of the electron at the bottom of the band. Application of the uncertainty principle to this type of internal transition – $\Delta E \cdot \Delta t \sim h$ – shows that the level from which the electron is missing must become broadened; in fact, calculations assuming the 'free electron' model indicate that its width under these conditions may reach values of the order of 1 eV. The degree of broadening will clearly depend on the position of the hole, the closer it is to the bottom of the band the larger the width of the level; such an internal change is reflected in the experimental distribution as a 'tail' which extends beyond the theoretical zero, taken as the bottom of the valence band.

The use of more sophisticated and sensitive detecting equipment has, so far, revealed the existence of 'plasmon' satellites in the low energy tail of Be, Mg, Al, Na; their intensity is, however, only of the order of 1% of that of the parent band and their position is roughly shown in *Figures 5.2* and *5.3.* They are so named because the interaction between energy quanta following transitions, with the plasma oscillations in the metal (produced by the electron bombardment) results in the liberation of a 'modified' soft X-ray photon; this, in the process of emission, yields up part of its energy to the collective electron oscillations of the plasma, and is therefore found on the low energy side of the main edge. In Mg, for example, it is found some 10 V below the Fermi edge.

As regards the intensity ratios, these should follow in proportion to the electron population of the L_1, L_2, L_3 levels respectively, and these, in a full shell, are in the ratios of 1 : 1 : 2. The experimental results show a very different picture, for the L_1 bands are extremely weak and have an intensity of less than 1% of that in the main L_{23} band. This arises from the filling of 'holes' in the L_1 band by electrons from L_3 in internal radiationless transitions involving an Auger process, and the absence of 'holes' available for *allowed* transitions from the valence band. Such a mechanism naturally shortens the lifetime of the 'hole', and we should therefore expect a broadening of the L_1 level; this is marked in the case of Al, for it is 2 eV wide. Since the width of the L_3 emission edge is of the order of 0.01 eV, little soft X-radiation can come from the L_1 band.

The same mechanism of Auger production can be invoked to account for the low intensity of L_2 spectra, as compared with L_3, for metals. Here the excited states in L_2 are filled by electrons from L_3 with the emission of an Auger electron from the valence band; there are thus no 'holes' available for accommodating electrons from this valence band, and therefore no emission of quanta in allowed transitions; for this reason the 'natural' expectation of a 2 : 1 ratio becomes 50 : 1 for Na, 20 : 1 for Mg, and 9 : 1 for Al.

In the case of semi-conductors and insulators the existence of the forbidden energy gap of the order of several eV prevents any Auger electrons which might result from an L_3/L_2 transition from finding a 'home', since their energy is only of the order of 1 eV; the L_3/L_2 ratio of intensities for such solids is found to be close to the theoretical value of 2. One might visualise the possibility of an Auger electron having enough energy to bridge the gap of, say, 7 eV; it could get this if an L_3/L_1 transition were possible in these substances: such an event would mask and hence annul any contribution which the L_1 excited state might make to the main valence band.

For completeness it is of interest to mention the existence of 'satellite' bands on the high energy side of the main band. Their intensity is generally small compared to that in the main band, though they have sharp emission edges and widths of the order of those found in the main band. They arise from transitions between valence and inner levels in atoms which have been ionised in more than one shell, either directly by electron bombardment or by an Auger process following on a first ionisation in an inner shell.

5.5 Absorption Spectra

Comparatively little work has been done on absorption spectra mainly because of difficulties of technique in preparing films of the required thickness, homogeneity and robustness. Such progress as has been made supplements the information yielded by a study of the emission spectra; it gives us an insight into the density of states and the form of their distribution for the unfilled levels above the top of the filled band. For absorption to occur the incident photon must have sufficient energy to raise an electron from the K, L, . . ., shells into vacant levels in the conduction band, leaving the atom with a vacancy, that is, a 'hole' in the shell concerned. The removal of this electron means that there is now an imbalance of charge; when this happens in a metal it results in an electron taking up a bound state just below the bottom of the conduction band; if it occurs in an insulator or semi-conductor it may give rise to a bound state just below the top of the valence band, as well as to a series of levels in the forbidden gap – these are the so-called 'exciton' levels which are closely spaced because of the high dielectric constant of the material.

The 'pair' combination of 'hole' in the valence band and electron in the gap constitute the 'exciton', and as each member of the pair is not wholly free, the states they occupy must correspond with non-conducting states in the material; however, as they are not tied to fixed locations such mobility as they possess enables them to transfer energy from one point to another in the crystal. Their presence is manifested by selective inflections in the absorption curve at certain wavelengths, i.e. for photon energies insufficient to enable electrons to pass directly from the valence band to the conduction band.

As with the emission spectra, we take selective elements and proceed to examine some salient features of their absorption spectra. Our final objective is the piecing together of the complete system of states in the bands and in order to do this, we use the fact that the absorption and emission edges must coincide for all true metals. For semi-conductors and insulators, on the other hand, the absorption edge should occur at a shorter wavelength, i.e. at a higher energy than the emission edge, and these deductions have been verified experimentally.

First the K absorption spectrum of the monovalent element Li which is shown in *Figure 5.6* will be considered. Here the K edge is not as

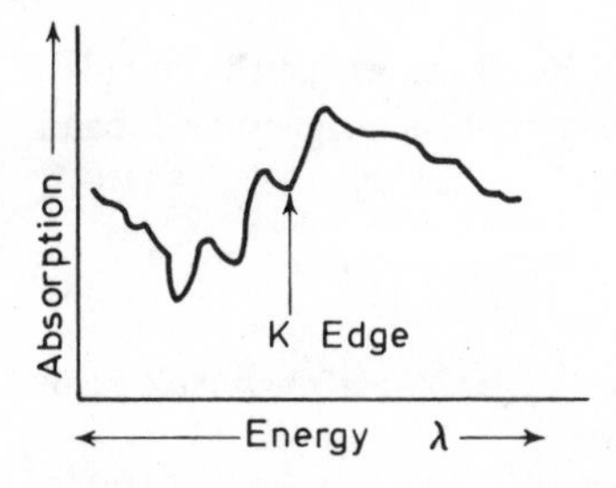

Figure 5.6 Li *absorption showing* K *edge*

sharp as one would expect for a metal, but it may, like the emission edge, be an effect of the crystal structure; however, the large fluctuating absorption effects on the high energy side of the edge seem to be a feature of the Li band. The complete curve of density of states, $N_p(E)$, obtained by superimposing the emission and absorption bands at the emission edge is shown in *Figure 5.7*.

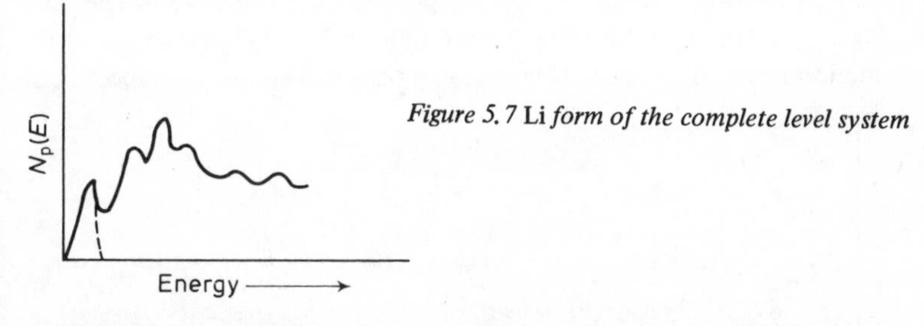

Figure 5.7 Li *form of the complete level system*

When we look at the absorption band of divalent Mg in *Figure 5.8* we see that the L_2L_3 edges are resolved; these are separated by about 0.25 eV on the original trace. Furthermore, the ratio of 'jumps' at the edge corresponds roughly with the required value of 1:2. The broad band on the high energy side is the result of L_1 absorption; this diffuseness of the edge is to be expected, since it has not been possible to observe the L_1 emission edge because of the Auger effect. When this curve

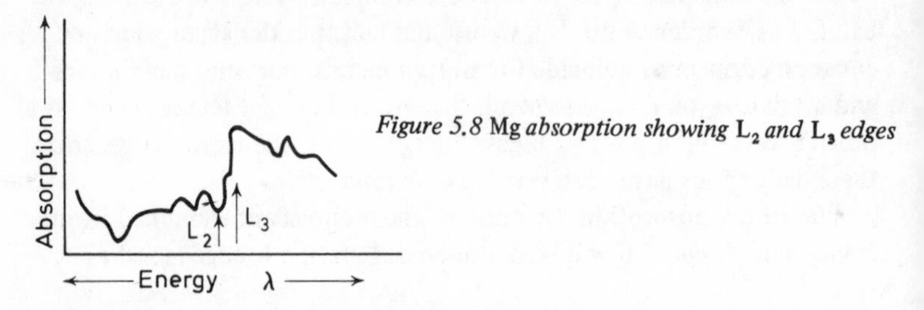

Figure 5.8 Mg *absorption showing* L_2 *and* L_3 *edges*

is superimposed on that for the emission spectrum, we get the complete picture of the energy density of (s + d) states of the magnesium L band system. Its form is shown in *Figure 5.9.*

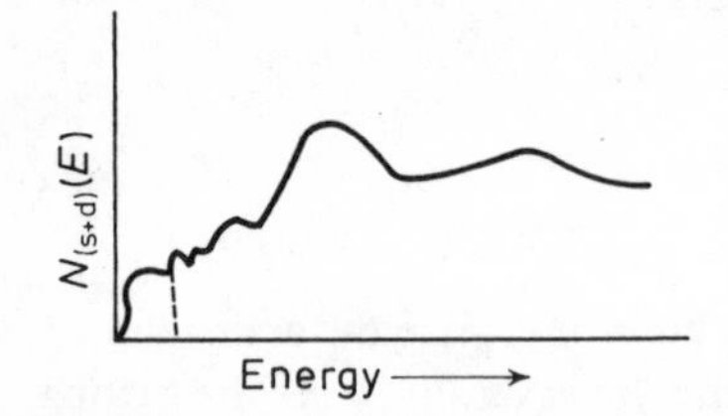

Figure 5.9 Mg *form of the complete level system*

The absorption in metallic Se is of much interest, for its interpretation involves the separation of the absorption in two separate levels. The electron configuration, $1s^2\ 2s^2\ 2p^6\ 3s^2\ 3p^6\ 3d^{10}\ 4s^2\ 4p^4$, shows that the M shell is completely filled; it comprises five levels. The main M_4M_5 absorption band consists of a doubly peaked curve sketched roughly in *Figure 5.10* to represent the separation of the M_4, M_5 levels of the conduction band. Since these M levels both possess d states, we expect them

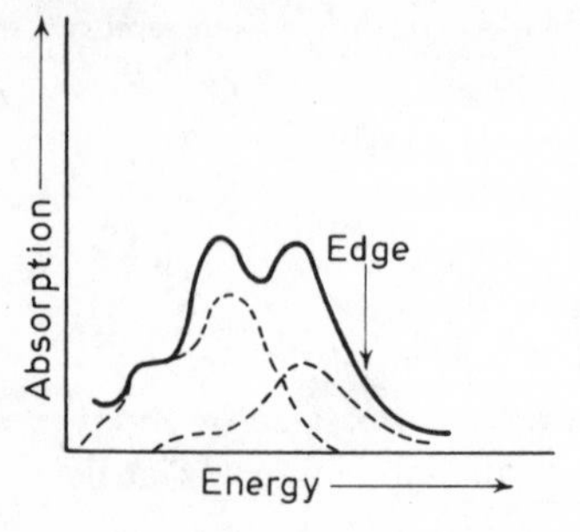

Figure 5.10 Metallic Se *absorption spectrum*

to give rise to bands of similar shape because the transitions involve higher p and f states; as a consequence their 'heights' on the trace should be in the ratio of 3:2, for this represents the probabilities of the transitions involved: their separation should correspond with the M_4M_5 energy difference. If we apply these considerations to the composite curve it is possible to resolve the main band into the two bands shown, which represent the absorption band for each level; here, we have somewhat different band widths, and a considerable difference in the density of states: the superposition of both must lead to the original main band.

Finally, it is instructive to compare the M absorption band in metallic iron with that of its oxide, Fe_2O_3, an insulator. The isolated atom of iron has its p sub-shell complete with its full quota of six electrons distributed among the M_2M_3 levels; the 'd' levels also contain six but could accommodate 10, so that there are empty levels available for p to d transitions even in the solid. As there is also some overlap with the N_1 levels, we are likely to get p → s transitions also. The form of the main M_2M_3 absorption band for both materials is shown in *Figure 5.11* where,

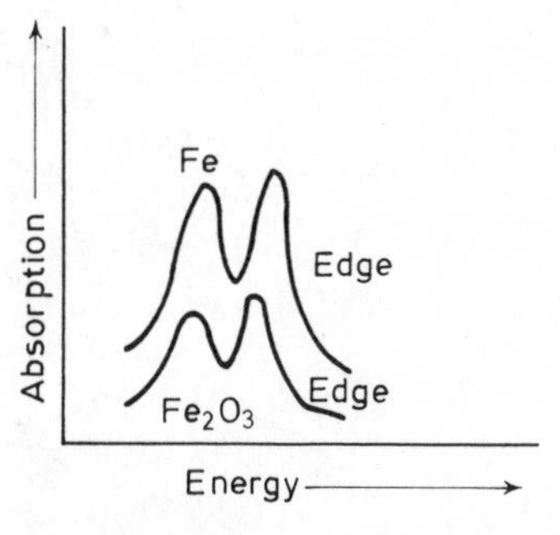

Figure 5.11 Form of absorption in metallic Fe *and its oxide* Fe_2O_3

for purposes of illustration, the displacement of the Fe_2O_3 edge has been greatly exaggerated; the doublet structure of the M_2M_3 level is clear, and the separation of about 1.7 eV gives that of the levels in the solid. A comparison with the oxide shows that its mean absorption edge is displaced some 7 Å towards shorter wavelengths, while the width of the band is also narrower than that in the pure metal. These observations are general in the case of pure metals and their oxides, and fit in with our notions of band structure of solids developed earlier, for the following reasons:

1. There is a 'gap' between the valence and conduction bands in the oxides.

2. The presence of oxygen ions in the lattice results in an increase in lattice spacing of the metallic ions; such an increased separation has the effect of reducing the overlapping wave functions and hence causes a narrowing of the band width.

Chapter Six

CURRENT PROBLEMS AND ADVANCES

6.1 Introduction

Having surveyed in outline the technique and some of the results of the soft X-ray method, it is desirable to assess what has so far been achieved in real terms in this field of research. We thus require to evaluate the validity of the experimental distributions in the light of the many subtle and complicating factors which can influence transitions between states and affect the shape and structure of the bands; further, we must compare the derived results with theoretical predictions given by the band theory. Let it be said at once that theory and experiment agree reasonably well for metals of low atomic number; the model of nearly 'free' electrons gives, in these cases, band widths which essentially verify the theoretical calculations, though the band shapes differ. Absorption spectra need a good deal more study from both angles, in order that their interpretation may be put on a sound basis. This chapter points out some of the experimental difficulties which, if not correctly assessed, lie in the way of agreement with theory, and the doubts they cast on the interpretation of shape and structural features. Finally, we take a glance at the new and exciting field of electron spectroscopy which has been developed within the past decade to provide a different approach for determining band characteristics.

6.2 The Recorded Curves

The purpose of the experimentalist is to ensure, as far as he can, that the recorded spectra are free from, or can be corrected for, a number of important errors inherent in the technique itself. Each of these is a problem in its own right. With emission spectra, he has to allow for the following:

1. Instrumental errors, causing broadening and reduction in resolution.
2. Surface contamination, arising from oxide layers and decomposition products of the beam, deposited on the surface of the target; these can alter the shape of the distribution and the slope of the emission edge.
3. Self-absorption in the target, an attenuation effect causing loss in the recorded intensity.
4. The presence of 'background', consisting of scattered and continuous radiation which affects the intensity in the region of interest.
5. The width and shape of the initial state, reflects on the measured band width.
6. 'Satellite' bands, their spread and intensities.
7. The method of excitation, which can affect band shape and cast doubt on structure.

In absorption work the researcher is faced in addition to most of these, with the necessity of producing really homogeneous foils and constant intensity radiation sources.

This is a formidable series of items which require consideration and are very difficult to assess correctly in any one set of experiments. Without the necessary corrections it is not possible to expect agreement with calculated band forms. Partially corrected or even non-corrected results are of value where relative shapes and values as between different types of materials are required.

6.3 Alloys and Compounds

An outstanding problem of interpretation exists when the recorded spectra of alloys and chemical compounds are reviewed. The theoretical band model throws little light on the understanding of the empirical results, for example the variation of the intensity distribution and the shift in energy of the maximum density of states. The former has been attributed to variations in the type and strength of bonding; the latter, to a redistribution of the total number of valence electrons, i.e. to changes in the proportion of electrons transferred from one element to the other in the compound. Both of these explanations are worth exploring further experimentally, with a view to developing a general theory of bonding. This could lead to reliable estimates of binding energies and give information about values in metal alloys.

In the case of amorphous semi-conductors or the covalent alloys formed from Groups IV, V and VI of the Periodic Table, where disorder arises from the random inter-atomic spacing, quantum theory indicates that the band structure characteristic of the crystalline state should be retained, but that a modification of the energy gap is to be expected. Indeed, photoconductivity and tunnelling experiments with these materials point to the presence of 'localised' states in the gap. These occur because the tails of the valence and conduction bands may overlap giving donor and acceptor states which are localised in the gap; the resulting mobility values are abnormally low.

It will be of great interest to see if the soft X-ray spectra of these materials confirm the tail overlapping and the modification in gap width.

6.4 Some Theoretical Problems

There are several fundamental problems with which the theoretician has to deal before he can present an accurate picture of band forms and account for the various structural features present in the records. It is as well to realise that the very existence of a 'core' vacancy, the end point of a transition, must reflect on the occupied states of the valence band, i.e. it automatically causes a perturbation in that band and falsifies the true distribution. Consideration must therefore be given to the problem of electron–hole interaction as well as to the existing electron–electron interaction and to what extent these affect the band shape.

In the approach of band theory one is confronted with the necessity of choosing the correct potential field for the metal studied, in order to arrive at the density of states for comparison with experimental results. This problem is made more difficult by the screening effect of the electrons on the field; its correct form should yield a value of the Fermi energy, and this is a sensitive index of the choice made. Although the experimental band shape is only qualitatively correct, it can give a lead to the calculations, and these can be made more accurate by referring to certain features in the bands, for example, the Brillouin zones.

It is generally agreed that a major difficulty in calculating accurate band shapes is posed by the variation of the transition probability P_t with energy in the band; though recognised, it has not been satisfactorily resolved. It is for this reason mainly, that the experimental distributions display a weakness, since the recorded densities are intrinsically related to the assumptions made about the variation of P_t with energy. Another problem on the theoretical side awaiting a solution is to explain why, in general, the emission bands are not sensitive to the crystal structure; formal band theory requires variation in band width as the inter-atomic spacing is varied.

6.5 Electron Spectroscopy

Within the past ten years, the soft X-ray method has been complemented by an entirely different technique for examining the energy states in the valence band. This is based on two separate approaches to the problem, (a) the study of the energy distribution of photoelectrons released when quanta of radiation of suitable energy, interact with electrons in the valence or conduction bands and (b) ion neutralisation at a surface, usually a metal, releasing secondary electrons from it and analysis of their energy distribution. The energies of the liberated electrons depend on the level to which they belong; for a given quantum energy those released from the lower levels will have less kinetic energy when they leave the atom than those which escape from higher levels in the valence band. The problem of resolving the energy distribution has depended on the development of refined measuring techniques for separating and recording very small energy differences. This has now been solved by (a) magnetic analysis and (b) retarding potential methods. The former has been mentioned in Chapter 1; it has been applied to the optical region with much success and gives precise separation of the different energy states of gaseous atoms and molecules, and information about the bonding and energies of vibrational and rotational levels. According to the Einstein photoelectric law, $h\nu = E_B + \frac{1}{2}mv^2$ where E_B is the binding energy of the electron in the level; a measurement of the kinetic energy of the photoelectrons for a given quantum energy of the incident radiation should give E_B for the various states. The fundamental requirements for the necessary precision are high resolving power and high signal to noise ratio. A schematic diagram to illustrate the principle of the magnetic analyser is shown in *Figure 6.1*. The whole apparatus is in high vacuum, and the sample is irradiated with quanta of known energy from an X-ray, ultra-violet, or optical monochromator (giving quanta of energy of order 1000 eV, 10 eV, 3 eV respectively). The resulting photoelectrons are put through the perpendicular semicircular focusing homogeneous magnetic field H, reaching the detector, usually

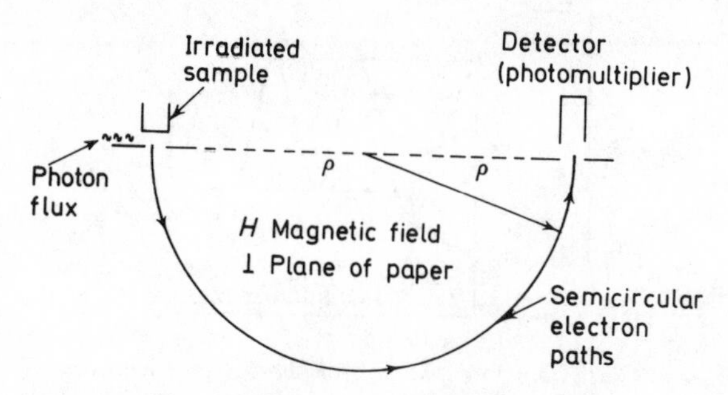

Figure 6.1 Magnetic spectrometer analysis

a photomultiplier, at the other end of a diameter. The radius of curvature ρ, is determined from $Hev = mv^2/\rho$, and the energy spectrum of the emitted electrons is determined by varying the magnetic field in small steps; the resolving power is given by $H/dH = v/dv = 2\rho/d\rho$, and is such that energy resolutions of the order of 0.01 eV are readily attained.

A second method of detection employs the retarding technique in high vacuum. The solid in the form of metal, semi-conductor or insulator is placed at the centre of a concentric spherical collector, and irradiated by quanta in the optical and ultra-violet regions through a special window in the sphere. A retarding potential applied between the irradiated specimen and the collector ensures that only those electrons with energy greater than the retarding potential can reach the collector. The form of the collector current for different retarding potentials is shown in *Figure 6.2.* This is clearly an integral curve, and in order to get the energy distribution, the slope has to be found at all points of the voltage axis. This can be done accurately by modulating the retarding field, i.e. by superimposing on it a small a.c. voltage (30 cycles, 20–100 mV) and measuring the a.c. component of the resulting current. If $N(V)$ represents the density of states per unit energy interval centred round V, the total current I, over the range of integration is proportional to $\int_{-V}^{\infty} N(V)dV$, hence the slope of the $(I - V_{RET})$ curve is proportional to the density of states. As the amplitude of the modulated signal is proportional to the slope at each point of the integrated curve, we have a ready means of displaying the density of states at each energy.

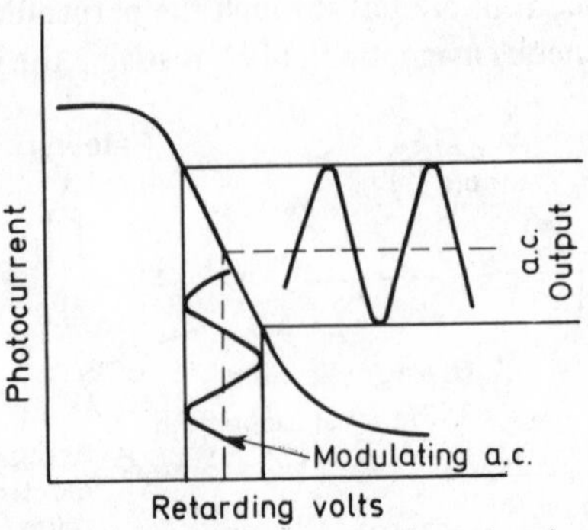

Figure 6.2 Retarding potential plot for photoelectric emission

Sophisticated electronic circuits are used in the actual recording of the energy distribution (*Figure 6.3*). In one set-up, the modulated retarding potential is slowly swept across the electrode system, and the resulting photocurrent detected by a specially designed photomultiplier and low noise amplifier in conjunction with a phase sensitive detector (*P.S.D.*). The latter responds only to inputs at the modulating frequency, and thus rejects noise inherent in the circuit. With photocurrents of the order of 10^{-14} A, the signal to noise ratio exceeds values of 100:1 and resolutions of about 0.01 eV have been obtained. The recorder traces out the form of the density of states to a base of applied retarding voltage; an example of the variations of density in the valence band of nickel is shown in *Figure 6.4* which shows a fairly sharp edge at the maximum electronic energy and overlapping of zones beginning about the middle of the band. Whether these structural features are real and represent a faithful picture of the system of states has yet to be decided unambiguously.

There are several factors affecting the accuracy of tracings quite apart from problems of design. In the first place, ultra high vacua and atomicall clean surfaces must be maintained throughout any 'run'. Secondly, attention has to be paid to the form of the field, so that a true velocity distribution can be measured, and allowances made for reverse photocurrents and contact potential effects between electrodes particularly affecting the low energy end of the distribution. Some further considerations involve the degree of penetration of the radiation in the specimen,

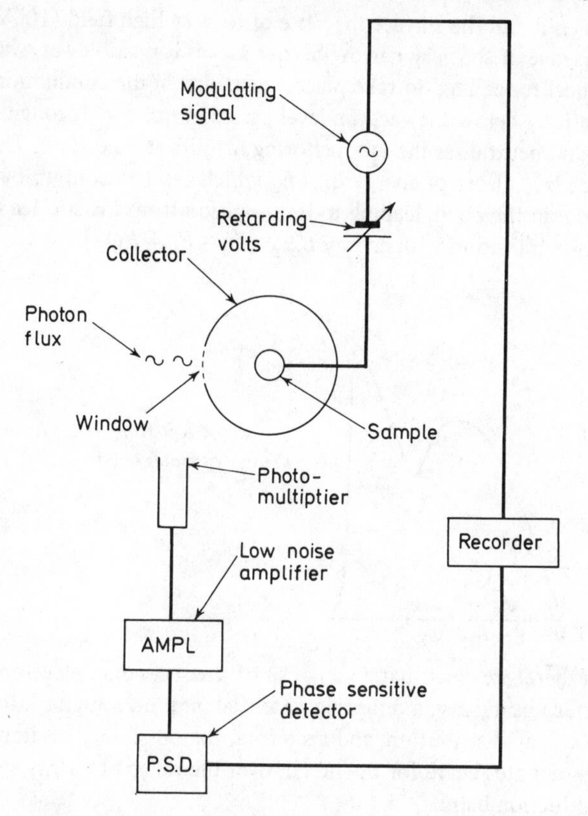

Figure 6.3 Energy analysis by retarding potential method

scattering lengths, interaction of emerging electrons with surface barriers and the question of escape probabilities.

Finally, it is instructive to consider the information yielded by the ion neutralisation process at the surface of a metal. It is, of course, well known that secondary electrons are released from metal surfaces bombarded by gas ions, and, in fact, this effect at the cathode of a gas discharge furnishes an explanation of the self-sustained discharge. The mechanism of the neutralisation process is illustrated in *Figure 6.5*, showing, on the left, the conduction band, Fermi level, and vacuum level for a metal. A positive gas ion (usually of the noble gases) is shown on the right, lying at an energy E_i below the vacuum level, since E_i is the ionisation energy of this atom; if its position is within a few angstroms of the surface, it will have the effect of creating a narrow barrier

B between it and the surface, because of its very high field (10^9 V/m). The presence of the very narrow barrier makes it possible for 'wave mechanical tunnelling' to take place; electron a in the conduction band at a depth E_a below the vacuum level passes in this way through the barrier and neutralises the ion, restoring it to its ground state. The sequence is a release of energy $E_i - E_a$ which can be taken up by electron b in the band, leading to its expulsion from the surface in an Auger-like transition with energy $E_{exp} = E_i - E_a - E_b$.

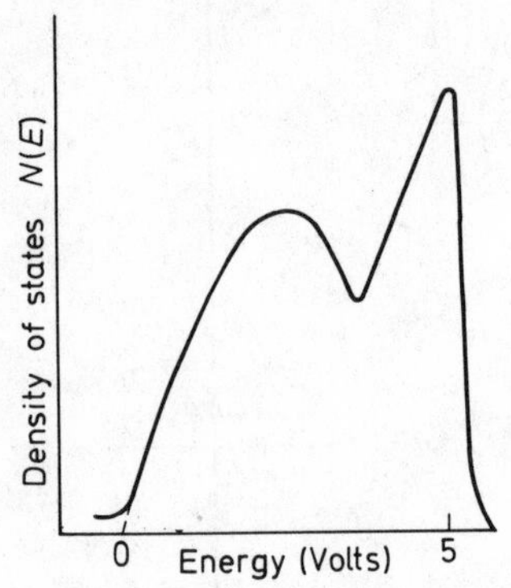

Figure 6.4 Retarding field: form of the density of states in Ni

It is therefore clear that an analysis of the secondary electron from the surface using, say, a retarding potential method can give information on the energy distribution, and therefore, the possibility of deriving the density of states curve for the metal, over the range of energy states in the conduction band.

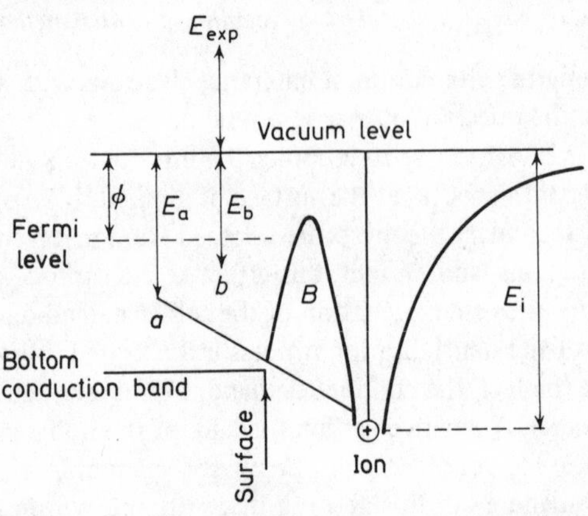

Figure 6.5 Ion neutralisation and secondary emission from metal surface

Experiment shows that the secondary emission ratio, i.e. the number of secondaries released per primary ion particle is independent of primary ion energy; generally, low ion energies, of the order of 10 V have been used in this work. The important experimental parameters affecting sensitivity are high ionisation potential of the gas atom and low work function of the surface. As with other methods, there are a number of important corrections to be applied, for example, escape probability through the surface barrier and transition probability variations depending on the initial levels. Such curves as have been published show many features in common with the results obtained by the soft X-ray method.

In conclusion, it would be correct to say that the opening-up of the field of the spectroscopy of the solid state, particularly by the recent advances outlined above, should give valuable information about the nature of the bonding in all classes of solids, organic as well as inorganic. The problems which confront both the experimentalist and theoretician are now recognised and to some extent assessed. There is every prospect that in the next decade we are likely to see a flowering of our knowledge and a deepened understanding of the properties and behaviour of matter in the solid state.

BIBLIOGRAPHY

Most of the material dealing with the soft X-ray method and its applications is to be found scattered through the literature of the past 25 years. Consequently, it is necessary to select a few of the really basic papers for more details than are presented in the text of this book. Good background reading for a first course would be provided by the following:

Chapter 1

Semat, R., *Introduction to Atomic and Nuclear Physics,* 4th edn., Chapman and Hall, London (1963).

Alonso, M. and Finn, E. J., *Fundamental University Physics,* Vol. 3, Addison–Wesley, London (1968).

Chapter 2

Rice, F. O. and Teller, E., *The Structure of Matter,* Wiley, London (1961).

Pauling, L., *The Nature of the Chemical Bond,* 3rd edn., Cornell Press, New York (1960).

Chapter 3

Shockley, W., *Electrons and Holes in Semiconductors,* Van Nostrand–Reinhold, New York, Ch. 5 (1950).

Slater, J. C., *Quantum Theory of Molecules and Solids,* Vol. 2, McGraw–Hill, New York (1965).

Raimes, S., *The Wave Mechanics of Electrons in Metals,* North-Holland, Amsterdam (1963)

Chapter 4

Tamboulian, D. H., *Handbuch der Physik,* Vol. 30, Springer, Berlin (1957).

Skinner, H. W. B., *Rep. Prog. Phys. Soc. Lond.*, **5**, 257 (1939).
Piore, E. R., Harvey, G. G., Gyorgy, E. M. and Kingston, R. H., *R.S.I.*, **23**, 8 (1952).
Cauchois, Y., *X-ray Spectra and Electronic Structure*, Gauthier Villars, Paris (1948).

Chapter 5

Skinner, H. W. B., *Phil. Trans. Roy. Soc. (A)*. **239**, 95 (1940).
O'Bryan, H. M. and Skinner, H. W. B., *Proc. Roy. Soc. (A)*. **176**, 229 (1940).
Skinner, H. W. B. and Johnston, J. E., *Proc. Roy. Soc. (A)*. **161**, 420 (1937).
Parratt, L. G., *Rev. Mod. Phys.*, **31**, 616 (1959).
Holliday, J. E., in *Handbook of X-rays* (Ed. E. F. Kaelbe), McGraw–Hill, New York, Ch. 38 (1967).
Holliday, J. E., in *Soft X-ray Band Spectra* (Ed. D. J. Fabian), Academic Press, London (1968).
Fabian, D. J., Watson, L. Y. and Marshall, C. A. W., 'Soft X-ray Spectroscopy and the Electronic Structure of Solids', *Rep. Prog. Phys. Lond.*, **34**, 601 (1971).

Chapter 6

Optical Properties and Electronic Structure of Metals and Alloys, (Ed. F. Abeles), North-Holland, Amsterdam (1966).
Siegbahn, K., *E.S.C.A. Applied to Free Molecules*, North-Holland, Amsterdam (1970).
Price, W. C., 'Photoelectric Spectroscopy', *Bull. Inst. Phys.*, February, 87 (1972).
Spicer, W. E. and Berglund, C. N., 'Photoelectric Energy Distribution', *R.S.I.*, **36**, 1665 (1964).
McEvoy, A. J. and Williams, R. H., 'Photoelectric Energy Analysis', *J. Phys. (E)*, **14**, 446 (1971).

INDEX